Why I rejected evolutionary theory

Aisha Mohammad

Copyright © 2018 Aisha Mohammad

All rights reserved.

ISBN: 9781719896511

DEDICATION

In dedication to my father and mother, and my sister and brother, to all those who have taught me and influenced me in learning, and to my brothers and sisters in faith, to all those looking for truth and open to finding it anyhow and any way and to those who are interested to know more about Quran and Islam

Aisha Mohammad

CONTENTS

	Acknowledgments	I
1	Introduction	p.8
2	What is the problem exactly?	P.13
3	Evolutionary theory	P27
4	History and Origin of the theory	P39
5	Problems with the theory	P47
6	What the theory means for people and society	P57
7	What the theory implies for humanity	P61
8	How evolutionary theory has been influencing all areas of life	P69
9	Criticisms	P82
10	How the Human body resembles the earth	p.89
11	The Soul or Spiritual self	p.94
12	The History and origin of humanity according to Islam	p.97
13	Conclusion	p.102
14	References	p.106

ACKNOWLEDGMENTS

With thanks to the library and librarians who gave me of their time and support, and special thanks to all my teachers and those who taught me, and to my friends who encouraged me, and to those who read over this work, and to my mother and family who supported me although not of the Muslim faith

1
Introduction:

I decided to write this book because of seeing that evolutionary theory is a theory taught in schools as a proven fact and has influenced much of modern day society and is generally accepted by people as something proven beyond doubt, but few have put efforts in to intellectually debate the theory, and it is largely been taken for granted, accepted based on scientific findings and so called facts, but they are interpretations. Can anyone prove the theory scientifically? I hope to argue against the theory on an intellectual level. This is because it is still a theory that has not been proven on a scientific level, so there is no real need to argue on a scientific level. Scientific information, also need intellectual interpretation. Data can be interpreted differently from one person to another, according to their understanding and personal beliefs. No one has seen the process of evolution or the in between stages or any sign of change or improvement in what has come before and we see all babies born as babies. In fact the intricacies of creation, human beings and the world around us indicate that their perfectness points towards their creation as they have been put in order, and the world as a whole has a governor, if it did not have how could all things be kept in order, and what makes the day overtake the night and the night the day, and what makes the sunrise in the morning and set in the evening? If there was no one governing the sun it could rise and set as it pleases. And how can we know something to be true? Is not science about interpretation of knowledge we gain through our senses? If so, is it not possible for all people to be scientific and analyse and interpret what they know, see and perceive through their senses and arrive to conclusions about the world around us, what is true and what is not, rather than being led to believe things by "scientists" who may not interpret evidence correctly, or who have an agenda (and how many in today's world have an agenda to keep religion as far away as possible) it has become more like a religion in protecting certain ideas and creeds and in not allowing those that threaten them, whereas science itself is merely a way to reach knowledge or truth, through trials, tests or interpretations of sensual data. In the past there were many priests bishops and religious people who were also scientists, as there were in the Muslim world also, in Arab Spain or Andalus. Science itself has no creed, but it depended on the interpretations of people who have their own creeds. In the present time it is atheism that is preventing anyone else from coming out with any other interpretations that show there to be a God. This is also not really fair since throughout history the vast majority of people have been believers in God and many held religions, even Socrates and Plato held notions of another life as well as this life and in their time there were atheists also, so that is not something new. However, there have been many individuals who have come to belief in God through looking at the world and studying it, or through some other research. Shall we say they are all unscientific and misguided people, seeing patterns in nature and its intricacies and its harmonies? Or shall we say that people who think deeply and reach the conclusion that we are not in this world for nothing and that there is a purpose to our existence, and that the finiteness of the world, the universe and ourselves indicates that we came from somewhere and shall go somewhere else after death. Shall we say such people are deluded and misguided, and those who do know, and cannot explain why we are here, and don't see any meaning in our births and deaths, are the intellectual and scientific, learned ones.

After this, what affect has this belief or perspective on society? I presume, all religions forbid lying, cheating, treachery and so forth, and teach people to behave well with themselves and others. In Islam a basic principle in allowing or forbidding certain things is whether or not something is harmful or beneficial, what is beneficial is allowed and encouraged, and what is harmful, to oneself or society, forbidden. In the end, the One who created everything knows what is beneficial and what is harmful. A human being in limited and their knowledge and information is not complete to make judgements and laws affecting many people. From what I perceive now, people are told they are now "free" to decide and do as they please – within some limits. Now we see crime as we did not see it before, drinking and drugs lead to many crimes, the breakdown of marriage, which is now seen as a religious thing which is no longer needed or something not so important has led to much heartache, cheating, distrust, illnesses, fatherless children and a lack of rights for women in particular and lack of rights for children who have the right to grow up in a home with two loving parents of the opposite sex – as that is how they came into the world, and a woman cannot take the role or a man or a man a woman – and know their lineage and ancestors, which unfortunately many do not now and it is not looked upon as important. All children have the right to know their parents and they have the right to grow up in a loving atmosphere, that will make them more productive and better members of society later on. Children who don't take the benefits of a strongly built and loving family in their early years may be effected later on in life in terms of productivity and relationships. So is religion so damaging and harmful then? And is the chaos we are now in with children who don't know their fathers or both parents, or the spread of diseases, or the increase in lying, cheating, and killing, which are all forbidden by all religions, is this chaos a better alternative, or a return to a natural state? Indeed, don't many people, atheists included claim to feel the existence of God, or to pray in times or extreme need- and see their prayers answered, are not these evidences that there is something in human beings that leads to religion, belief in God and so forth, and that in turn leads, in many cases to better behaviour. In Islam, a person is taught to be good to their neighbour, whoever they are, and not to harm others, and to be good to parents, to help the poor and needy and be good to children, their families, their wives, and all people and not to kill or harm others as claimed.

So I hope to open the minds of people who have perhaps been closed and covered by evolutionary theory and other claims, and I hope the reader will read and learn at least to think for themselves, I too learned evolutionary theory at school and never thought to ponder it until I met with some Muslim people who told me that they too learned it in school but did not accept it as a proven fact. We were told we must accept it as proven, so it never occurred to me to question it, when I did however, I began gradually – over some time (it is not easy to reject something you have been just like that) to realise that the theory leaves more questions than answers and does not explain anything in essence. I have realised from being told many things and taught many things that I later found not to be correct, so I realised the need try to depend on my intellect, and understanding and sensual perception, not on other peoples claims. And claims about where we come from, why we are here and where we will go are all big questions we need answers too, and these ideas affect society and our activities within it, and between each other, and what if the claims that there is a judgement day are true and they are just ignored? Is that not something very dangerous? We have been told to be busy, not to think, or use our brains, just follow the claims, but where are they heading? Do you know who you are following? Is that not the life of a cow? I guess Nietzche said something similar. So I hope what I right may intrigue or influence some people at least who are still open to thinking and learning and that it may be of a general benefit to readers.

Aisha Mohammad

2
WHAT IS THE PROBLEM EXACTLY:

What are the reasons for so many of the problems in society? Why is there so much depression, anxiety and mental health issues amongst people in society? Why are so many people confused about their reason for existence and where their lives are heading. There is so much dissatisfaction, despite all the comfort and ease available to people, and despite all the availability of knowledge, and advances in science, people still seem to have more questions than answers. The problem is that many people in society are taught wrongly that they are the great grandson of an ape. This theory holds that human beings have not been created, rather they 'developed' out of lower life forms. It is a theory upheld by many in the scientific community, and is hailed by atheists as a proven fact. However, we hope to show that it is not. That is easy, since a proven fact requires acceptable proof and there seems to be none of that, only imagination and stories. Apparently it began in the water, according to this theory, fish (which normally cannot live outside the water without dying) over thousands of years developed the ability to breath outside water; they also apparently developed legs and many other things and walked out of the sea. Walking not being enough, some of them developed wings and flew away, leaving some 'in between' stages.

They also hold that all life forms – including human beings developed in the same way. When objects that such phenomena are known to be scientifically impossible, and is an illogical explanation of life, the answer is given that it took thousands of years, and that is why we do not witness such a process. The problem is that if something is illogical, whether they argue it took one day or many centuries, if it is not possible, it is not possible, and furthermore needs proof to back up that it is not an imaginary story but a proven fact.

Although it may sound like an appealing and interesting theory to some, it only makes an attempt to explain human origins outwardly – inwardly it is contradictory and it leaves out many details, such as why we are on the earth in the first place. How could the earth evolve itself out of nothing and then go on to design itself? Is that not making the earth like a God? The earth did not evolve out of gas. Actually, it is generally held that the earth had a sudden beginning – a big bang. Yet, as we have mentioned, a bang is usually caused by an explosion, which itself has a cause outside itself. A person sets a bomb off, or bangs a drum, or something falls. In such instances we see contrary to the order that such persons claim. The claim is that after this, probably life on earth came later but in the amount the time it would have taken for this slow 'evolution' the sun would have burnt out and the earth would probably have dried up by then. And since the world itself goes through its own changes, by the time people or animals evolved to suit a particular climate, as they believe, such a climate would itself have changed yet again, and they would need to evolve themselves to suit the new climate. Not only that, but the fact is that animals and people move around and do not stay in one place. We see people come from Africa to live in England, and we do not see them evolve to suit the British climate. In the time that they claim it would take for such a process to happen, it would not be any use to those who moved to Britain, and in fact their offspring may move again to another land, which means that adapting to such a place would not be of any use to such persons. Life relies on being fully able to eat, see, experience and so forth, so how can it be claimed that creatures had to spend time on earth with no ability to breath – until their lungs developed, or half developed eyes, or a stomach not yet developed so that they could not eat, what then would be their life? And what about the cell, did that evolve too? When one thinks about it this way, one can only envision that all creatures were in need of having whole organs to use – such as eyes and so forth. We can further elaborate on the fact that eyes, were designed for seeing, and mouths for eating, that is, half evolved eyes or breathing ability just would not benefit the creature. Similarly now we do not see evolution taking place now at all. When we look around we see people, of different colours and nationalities, different social classes and genders that go about things differently and appear different but are essentially equal and have similar abilities in life. Society just would not be able to function with some more evolved people than others. So what happened did the first people who supposed to have evolved wait for everyone else, to make sure they could all live equally. That was nice of them, how considerate. But who evolved first according to this theory, was it a man or a woman? How did they reproduce? Was it a black or white individual? Does it matter? Well if you are considering equality, the Nazis did use this theory to suppress black people who they considered to be less "evolved" than themselves (such as a lower life form) because it was embedded in the theory itself. When we see babies born we see them born the same as any other human being, there is no evidence at all that they are slowly developing into anything else, and studies of genetics, demonstrates that babies take their traits from their parents, or from some generations back, rather than from the world around them, which is what the theory seems to suggest. As far back as we know we find human babies growing into human adults. Since recessive genes bring forth things that go back far in our genes we might expect that every now and then we see a monkey baby being born, or a baby with hair and big teeth, or perhaps a baby swinging from a tree scratching its armpit, at least that would confirm that they were our ancestors and not some other species, but that does not happen. From observing humanity, one can only really see humanity as having one source and that, as all children as far back as we see come from a man and a woman, so did all the humans, except the first two who were created. And the evolutionist claims that mutations cause changes in the species, has anyone ever seen such a thing? We see cows with five legs on occasion, but that cow will still mate with a four legged cow and give birth to another four legged cow, because basically mutations are not passed on. Mutations do not cause things useful to the species, such as increased intelligence, but rather, things such as the five legged cow (and the cow does not walk on its fifth leg, which is attached to its back). Each species has its own gene pool. So if there is no evidence of evolution from looking at the human species itself and no evidence from the fossil record – because they have never been able to find these in-between-stage-people, they conveniently died out somehow (apparently) and left no trace they were ever there, then there is no evidence for this supposed evolution, and it's just an imaginary scenario thought up by people who have something against religion. In fact evolutionists themselves use the term 'imagine' when elaborating on how events took place, because, that is all one can do without evidence, and when something happened in a time when no one was around to observe how events took place. Another point is that, as we mentioned, science, traditionally, has placed great importance, onto what can be observed and seen, if we cannot observe or see evolution, then we really cannot accept it as a scientific theory. Another thing is that if we were the last creatures to have evolved on earth, we should really be less in number than other creatures, yet the earth is very populated by people, demonstrating that we have been around for a very long time – the same time as other life forms.

Evolutionary theory also does not explain why we human beings have religion. In all regions of the world there are people who believe in God, the vast majority of humanity professes to believe in a God, or a religion, even if slightly different. How could we hold such a view if we are a result of only random mutations? Neither does it explain why we are so unique that we are the only species on earth, who cover ourselves with clothes (if we were previously monkeys why did we lose our hair to cover ourselves with clothes), we build schools, universities, hospitals, we have developed technology, we have been into space (or sent others into space), we use planes and so forth, other animals do not have these abilities. Animals do not have languages (rather they have sounds that are understood between them), law and law courts and police to keep social order, they do not write books, create computers, establish scientific institutions for research about the earth and do not wonder about anything more than what they eat daily. Human beings are obviously quite different – just focusing on the fact that there are so many digits similar with apes does not change that, or mean that we came from them. Even physically, we walk on legs –we have long legs and small arms, they have long arms and short legs, we have different colours of skin, hair and eyes, monkeys – apes are just brown colour with brown eyes. Their eyebrows are protruding – ours are not, their eyes are like marbles, ours are different shapes. They have a large protruding nose, on a base like a coconut, we do not have that and their mouth is large, wide and protruding, with huge teeth. They are fast in the trees and slow on the ground, we are the opposite, with our long legs and short arms designed for walking on the earth, while they have long arms and short legs, designed for swinging through the trees. We have culture and social cues, we give people things rather than pick flees out their fur for a favour. All of the above demonstrate differences between monkeys and humans. Such a belief also does not explain where we have come from in origin – we clearly have a beginning - if we were to go back through all the various generations of humans who have passed away (which probably we would not be able to do in our life times) we would get back to the first of the humans and there would be a time before humanity existed, and in fact, before everyone existed. We therefore, have to have come from somewhere. Everyone on earth dies – there is nothing we can do about that, no matter how sad we are, how much we dislike death, how many vigils we hold for groups of individuals who die in ways we deem unfair we cannot prevent death for anyone or bring anyone back. It will come to all of us. Sometimes doctors can keep someone going a little longer than they would otherwise live, but they will inevitably die as all humans do. Some of us die in our sleep in old age, some of us die from diseases, some from accidents – such as road accidents of work related, and some are purposefully killed as in war or murder. There are some who just drop dead and people can find no explanation for their death. Some die very young and some very old – if one reads the newspaper one will realise how many people die at a younger age rather than an older age. One will also notice that the majority of people are not old – quite a few have died in youth. Therefore, thinking about death and what happens thereafter is one of the most important questions one will have to ask oneself, in this life. It is a question you need to find a truthful, honest and satisfying answer to before all others – such as careers, marriage and so forth. It may be that death could come before you have achieved anything, and if not, at least you will know where you are heading.

Why do some people doubt about God if He is really there one might ask. Individuals usually doubt due to confusion about scientific theories and findings or due to lack of experience or understanding. However, if a person where to wake up in a room, or a particular area of London, when they had previously been abiding elsewhere, certain questions would come to one's mind automatically, such as "Where am I?" "How did I get here?" "Why did I end up in this place?" "Where do I go now? What do I do now? What will become of me?" In the same way as asking such questions are natural in such circumstances, they are even more reasonable when asked about ones existence on the earth. If someone were to walk down the street and see an old fast-food box of fried chicken, one might rightly assume that another individual has been there, ate it and thrown it down – even though you have not seen them, they have left their trace. One knows that a box of chicken bones has not eaten itself, or thrown itself down. Also, if one hears loud booming music coming from the street, one knows that there is a person playing that music, probably from a car. One knows that such sound has not appeared by itself but has someone instigating it from where you cannot see. When one considers these small issues – the chicken box that has obviously been thrown down, the loud music and so forth, one must look around at the world and universe at large, it is so big and so vast and yet there seems to be a perfect order in its existence, there are so many details and so much intricacy, is there any of humanity that knows its details? We see scholars dedicating their lives to learning about a particular thing – such as plant biology for example, but can such an individual tell you all there is to know about his or her subject (let alone where they come from)? Even with their PhD or can they know all there is to know about a blade of grass for example? What about the origin of this blade of grass? One may wonder what my point is and why one would be interested in a blade of grass. I am simply trying to show how something so small and detailed could not come about by itself and that there could not be an accident leading to such a creation. Even when one looks at oneself, how would we function with four fingers instead of five? Or if we only had one eye rather than two, or if our ears were on the tops of our head rather than at the side, one should realise that we clearly have been designed to function in the best of ways, and throughout all creation there is an order and purpose. Even though we have not seen the universe being created, we can know that it did not create itself. The sound human intellect simply would not accept that something could just appear or just be there and science itself, in its 'big bang' theory has confirmed that there was once a time when the world did not exist, and, if something did not exist at some point it must come from somewhere when it came into existence. We see the world in an order, we see that plants grow from the ground, that fish swim in the sea and eat certain foods that the sun rises and sets at certain times that we can predict. If fact, science is really the study of the order of nature, trying to work out and predict how our world works. One who studies the laws of nature, is therefore adhering to science. If there were no Creator, and the world was just there, firstly it should be eternal, and yet it has a beginning, secondly we find that it is orderly and detailed and that we can predict things such as the sun rising in the east each day and setting in the west. If there was not a power over the sun controlling it, then there is nothing to prevent the sun from rising in the west some days and setting in the east on other days, or not rising at all some days or setting mid-day and rising in the middle of the night. We would not be able to predict this occurrence. Perhaps the sun would even go off course and rise and set on Saturn instead of the earth – that is possible if there is nothing controlling its movements. We cannot see gravity, yet we know it exists because if we take an object and drop it we know that it will fall. We see its outcome, but what is it that keeps gravity in order, so that things do not just float away into space? If there is nothing controlling this earth then what if gravity was to cease and everything were to float away. We see that everything on the earth has an order and that it follows clear laws so that we can predict actions, such as the movement of the sun and the moon. The fact that changes occur and that life is not static further proves the existence of a Creator. The earth itself is evidence of a Creator and people who claim that there is no evidence are either confused about what can be classified as evidence (they are trying to apply laws of created things to the creator) or they are just refusing to look and use their intellect. Since our Creator is not within his creation and we cannot see Him, we can only know Him through meditation and reflection on the earth we can see. We see his Power everywhere in the world and everything in the world proves His existence. Belief in God is also something that mainly concerns the intellect, rather than the senses.

When one uses ones intellect one will see that we were created by One who is far different from any life on earth. It is the case that the painter of a picture is entirely different to the painting that they have painted, while having full knowledge of every infinite detail in the painting. They are not inside their painting, but everything that they do is from them. They therefore, have full power over it, to change whatever they want. If they want to destroy it altogether, they can do so, if they want to improve it, they can do what they like to it. Our Creator is entirely different to us, and since He created us He has full power over us and knowledge of everything that we do – including what is in the breasts of men. If we look around the earth we find that the majority of people believe in One God, although they may have slight differences or deviations, because God almighty instilled within the human being the ability to know about him. He created the human intellect to know him and the capacity to understand, the soul and conscience to be good and righteous, the feelings and sentiments to be kind and humane. Some people fail to notice the mercy of God, taking it for granted. The fact that so many people testify to 'feeling' or experiencing a connection with God in prayer or meditation is another evidence for the existence of the Creator. For example, some of us in England have never been to Africa, we have heard about it a great many times, however, from a lot of people and it has been drawn on maps. Therefore, because we know of a great number of people who have come from there or have visited there and experienced it, we do not become sceptical and disbelieve them, we know that a lot of peoples experience of something at different times, and perhaps in different places, cannot be a mere coincidence or fluke.

Through our reasoning we can also argue for a Creator with the following points. Every nation needs a leader to keep the society in order, every school has a principle, every work has a producer, and every work of art has an artist behind it. For example, it would be absurd (and no sane person would believe it) to say that this book came into existence by itself and that it is just there, that there was no writer behind it. Without writing my name or explaining to you, everyone would know that a person with a will and who put and gathered information from different sources and put it together and organised it wrote this book. A potter formed a clay pot – you cannot leave clay and expect it to change into a pot. According to the laws we have on earth, everything must come from somewhere, people or things do not come about by themselves that should be obvious to everyone. Nothing comes into being on its own and no family or state can function without a responsible head. There must be a stronger Power governing our world and keeping the earthly laws in place. If people came to earth by chance, our entire life is based on chance and so the whole of human existence is meaningless. A sensible person would conceive of his life, and the things that he has done in life, as meaningful and a rational person would not leave his existence at the mercy of fluctuating chance. Every reasonable person attempts to give his life meaning and purpose and set himself a model according to some design. Individuals, nations, groups and courses all plan carefully to produce desired effects. Since humanity is only a small per cent of the universe and can plan and appreciate planning, then his own existence and survival of the universe must also be based on planning. The second point is that it is illogical as we have mentioned, to say that our earth or anything else could have come into existence by itself, since that contradicts the laws of nature that we already know. It is not eternal – nothing on earth is eternal so it comes and will go somewhere. Thirdly, the laws and details and organisation of what is around us indicate that there is one looking after the earth, one who has power over it and us and who governs our affairs. A child may begin by coming to belief in

God's existence by wondering and asking, where was He? Children are often taught that God is in the sky. This is often from cartoons and various stories and claims of older people who don't know how to explain to children well. So there are those who leave childhood with the idea that God was like an old man, living behind a cloud. When one gets older they can see the absurdity. How could God be in the sky when He created the sky? Because if God created everything, then He created the universe and so could not be in the sky, which He created. The child may think to themselves that if He were in the sky, aeroplanes and birds would have seen Him , and so He could not be in the sky unless He is hiding behind a cloud. But the clouds move, so He could not stay behind a cloud unless He moved with the cloud. Then that perhaps He is invisible, but then it still does not explain the Creation of the universe or if God were just in one part of the sky – or in His creation at all, then how would He know about everything, or see everything she thought; if he is confined to one place. God cannot be in the sky. So perhaps God is in space. But if God is in space, why have not the astronauts and so forth seen God? And there are shooting stars and so forth, so she could not see how God could be in space. If He was invisible or in one place, it does not explain how the universe and sky and earth were created. Logically, the only conclusion to explain the creation of the heaven, earth and sky is the realisation that this could only happen if God was outside His creation. One cannot be within what one has created. Since God has created space and time, God

De-evolution from monkeys and a return to humanity:

is not within space and time; God is without a place and without being subject to time and other human limitations.

3

EVOLUTIONARY THEORY:

In school we were taught in religious education that there is God who created the world and Jesus peace be upon him, was his Prophet or some say "son" which was meaning at the time a person close to God and Adam and Eve were the first humans. We were told if someone leads a good life they will, if they are a good and devout believer, go to paradise and if not they will receive punishment. We then in science were taught that the reality is that people evolved from monkeys and that we are just a type of animal. We were given the idea that religious stories are all myths and nice stories to make people feel better and to give an explanation about the world for people who have no knowledge, and scientific stories are all facts. Whereas it is not advisable to accept and believe everything and every story we are told as there are many myths, we should be able to distinguish real myths from real events and stories And what makes something a fact if it taught to us from scientists and myth if it comes from clergy or religious persons?. This theory is largely passed off as fact in schools and in learning institutions in general and amongst scientists disbelieving in it is akin to heresy. This only seems to indicate for me that scientists are trying to turn their discipline into a religious order which it was not meant to be, and is it a bit hypocritical if they are claiming that religion and science do not co incide. This is their dogma. Atheists use it in their defence more than anything else. For example, Richard Dawkins (a rather infamous atheist whose books seem to be well known and highly thought of amongst the atheist elite) wrote a book entitled the 'God Delusion' as he calls it. I opened it one day expecting some deep and well thought out arguments that I may have to spend time going over. However, fortunately for me his arguments were rants and based largely his personal acceptance of evolutionary theory and did not bring any strong evidence. The fact is that for those who do not believe in evolutionary theory or natural selection, basing belief in God on that does not really apply to them or make sense when used as a critique of other ideologies. This theory has not been around for that long really, since it came about in the eighteenth century, so were all the great scientists that we mentioned, who believed in God before, as he says, stupid and ignorant? And only he and those who agree with him, he believes anyway, are the intellectuals? But it does make things easier for me now, since if atheism is largely based on this fictional theory, we only need to show that it never actually happened to those who believe it. I did actually believe in it before, since I, along with everyone else, was taught it in school. However, from considering it from all standpoints – intellectual, historical, 'scientific' and so forth, and from, again, using my God given senses, I can understand that it did not happen and why not. Firstly, let us look at whether there are similarities between monkeys and humans. I would like to point out, that even if there are this does not mean that we came from them, it only means that there are similarities, which can be expected when we are living together on the same planet. We both eat food from the earth and breathe the same air and would most likely have the same origin (which does not mean evolutionary). However, I do not see, as Darwin claimed to have seen, similarities between these hairy animals that swing through the trees and we who walk on land and who cannot swing through trees very well. You would have to remove all the hair, change their physical structure, their face and colouring. Even then we would not be identical. Darwin claimed that physically we are very similar to apes in our bone structure. I wonder how he came up with that since apes have long arms for getting through the trees and they have shorter legs. When we see apes walking they are slow and lumbered because it is not so natural for them, the same as a bird who walks sometimes – it is more natural and easier for them to fly. People may climb trees sometimes, but it is harder for us and we cannot swing through them as apes can. We walk on the ground. Genetics means that everyone inherits traits from their ancestors, yet we only inherit from human beings, apes are not our ancestors, so we don't inherit from them ape abilities, thick body hair and long arms. Babies, as far back as we can trace have always been born as human babies, there is never any sign of ape traits in them. Had we evolved why do we not see some of these 'recessive genes' that can be passed on from generations back influencing the face of the baby or making them skilled tree climbers. If they inherited these recessive genes, some babies should be born hairy, or with longer arms than legs, or with something else from a so called ape uncle. Babies do not show any signs that they were apes or that they are evolving into something more than human either. Therefore there is no sign in humanity that we were ever more (or less) than human. Genetics disprove it, because according to genetics genes are passed down, and can only come from those we are actually related to. We take genes from any type of human being – black, white and anyone in between because we are all humans and all have the same origin and abilities – there are no 'races,' just human beings. We do not inherit from apes because they are a different species and they only pass their genes on amongst one another. Things do not evolve to suit the environment, whatever environments we are in we are predominantly physically made up of genes.

Evolutionists say that the fossil record shows that life first appeared on earth at least 3.8 billion years ago and that evolutionists say that all complex life evolved from these first simple forms. They think that diverse life formed from single-celled organisms to mammals with complex anatomies, such as giant whales.

But why does this mean that life evolved and where does evolution come into this, since finding fossils by themselves does not mean that a people came from lower life forms. In fact there are many types and degrees of life on the earth, so why does it mean that finding fossils of more basic forms of life mean that higher forms come from them?

Life is the ability to take in and expend energy, to grow and change, to reproduce, to adapt to its environment, and in more complex living organisms – to communicate. The cell is the fundamental unit of life, capable of replicating itself and carrying out all living processes. Even the smallest of organisms have at least one cell, and almost every cell of every living organism has its outset of molecular instructions. Within each cell the threadlike chromosomes carry hereditary information in the form of genes, which are responsible for the particular characteristics of an organism. The set of instructions in a gene are mainly recorded in the form of a molecule called chromosomal deoxyribonucleic acid – DNA

The DNA of an organism carries information from one generation to another allowing certain characteristics to be passed on from parent to offspring.

Therefore, the cell, the smallest of units for and creature cannot be broken down, which means it also cannot have evolved because evolution relies on developing from something less developed and there is nothing less developed than the cell. Therefore, there must have been a point or time when the smallest of things were created and that they must have begun. How did the cell come into existence? How did it begin? When we know that there are some common reasonable laws to be followed in terms of scientific laws we know that each and everything relies of external factors for its existence. A fetus develops and comes about through the mother and father of the child, houses much be built by a person, a chicken must lay an egg, a person must write a book, cooked food must have had a cook, and therefore in the same manner, according to these same principles, a plant must have had a Creator, as should a tree, an apple, fruit, insects, animals, birds, mountains and so forth. It is not intellectually reasonable for anyone to assume that things came about by themselves, animals created themselves, creatures gave themselves wings or legs to move about and so forth as they chose. It also does not make sense since if fish had been able to do such things in the beginning, we should still be able to fly and develop as we wanted now, as flying is actually very beneficial, especially in traffic jams we should, if evolution was for our convenience, be able to fly when needed. Why would animals loose abilities they previously had which would have been of benefit, such as flying or movement through the trees, if a person had really developed, they should be able to retain all their previous abilities, while mastering new ones, if that were the case. The fact that there are many different types of animals, insects and creatures in general that don't have this obvious hierarchy indicates that they are not from a common ancestor, but are different types of creatures.

The huge array of life is divided in three domains or superkingdoms – Archaea, Bacteria and Eukaryota – which encompass all life forms from plants and fungi to animals. The first two domains are basic, the more advanced eukaryotes have a cell nucleus, which contains the cells genetic material, DNA. They range in shape and size from single celled, to complex, multi celled plants and animal

Therefore life as we know it is vast and in different domains, there may be some fundamental similarities between different types of creatures, and some more complex than others, or in degrees. However, this does not mean that there is a hierarchy and that some come from others. Similarities or hierarchies of anything do not mean descent and do not imply a common ancestor, because there is nothing to show directly that one such thing has come from another, rather it makes sense to say they share commonalities and a similar creation. It would not be absurd to say that they are evidence of One Creator.

A capacity to grow and repair is one of the key defining features of life. All organisms grow by increase in cell size and division

Being able to grow repair does not mean that a species of creation can merge and become another on account of small subtle changes. Rather, it is evidence that the species has some form of self-protection. Growing within a species is apparent and without doubt, however the claim that one species can gradually and slowly change and evolve into another species is the question of concern.

Evolutionists do not distinguish human beings as a creation distinct from animals, they say that human beings are a type of animal, and that they are primates, which they say are types of tree living animals, that all share a common heritage, such as Gorillas, Gibbons and so forth. They say the human being is different however, since humans, they claim, have developed the ability to walk on the ground on two legs. They say, however, that the human being shares other features with these primates, such as having five digits on their hands and feet, opposable thumbs, which can be brought into contact with the tips of the fingers other primates have opposable big toes as well.

They also have large forward facing eyes, which allow good depth perception. Nails, rather than claws on the fingertips, year round breeding and gestation periods with only one or two offspring produced per pregnancy and flexible behavior with a strong emphasis on learning.

However, when looking at what it has been said about the humans being as a type of primate, they are clearly not, since primates are classified primarily as tree living animals. Human beings are not great tree climbers and their bodies are clearly not designed for that purpose, for this reason, they have long legs – designed to help them to walk and get by on land. They have shorter arms since they are not designed for swinging through trees or pulling leaves and other things from trees. Therefore, even if there are found to be some similarities in other areas – which is not surprising since they are creatures that live on land like us, and we can find similarities in the human body to many things, such as plants, and indeed the earth. Therefore, some of these similarities do not make the human being from the same category as monkeys, for he is not, and definitely it does not indicate that he came and evolved from them.

In some books we find mention that humans are the cousins and closest relatives to chimpanzees. They say there are lesser apes, and greater apes, such as chimpanzees and Gorillas, and humans and their ancestors have been put into their own category, which they refer to as hominids. They then add that because humans share a lot in common with African apes, it is then better to group humans, apes, and chimpanzees together under the title of hominids. They add that humans are genetically closer to chimpanzees than gorillas, and they say they have been called the third chimpanzee.

This is one of the ways atheists try to twist science and evidence to bring humanity away from any understanding or interpretation that may coincide with religion. Since religion had previously held the human being to be a special creation, they have to come and put human beings amongst one of the more disrespected and undervalued types of animals, that people laugh at and we see in zoos, swinging through trees, eating bananas and screeching. Maybe they will ask us to use our imaginations, but still, even if there are some small similarities that can be found – or imagined between humans and chimpanzees, we can find and bring many differences that demonstrate that the human being is entirely different, and is his own species if we may use such a word. Having any similarity in structure does not indicate at all that one thing developed from another, as it does not indicate that if we see a fellow human being that resembles a member of our families, perhaps our sister, does it mean that she must be the relative of our sister, even if from another family, and perhaps even another country. Therefore, this sort of jumping to conclusions and drawing faulty conclusions from things that is not evidence is a bit worrying. They do however, acknowledge that the fact that chimpanzees walk on all fours and humans on two legs make their skeletal structure different.

Evolutionists further claim that fossils are evidence for this so called evolution, they claim that hominid findings began from East and central Africa with many finds from the Rift Valley. They claim that early man walked up-right while large brains and tool making came later, with the appearance of the Homo. They claim that findings from fossils of possible early hominids were found in East and central Africa, claiming to be more that 5 million years ago. They claim that fossils show that this was able to walk upright on legs as well as be able to climb trees. They claim that the fossil record shows that from 4.5 million years ago, a range of fossil species emerged, which were well adapted to walking upright, but did not have the long arms and large brains as the Homo genus. They further say that we are currently the only hominin species on the planet, and this is unusual, for most of human evolutionary history, there have been several species overlapping each other.

Firstly, just finding fossils of skeletons from years ago that may resemble humans as well as non-humans does not mean exactly that this is evidence of one species evolving into another. This is the problem here of interpretation, for personal beliefs, enthusiasm, hopes, dreams, and creed can all get in the way of interpreting such findings. Would someone who did not accept or believe in the theory of evolution accept such fossils as evidence? I don't think so. Perhaps they would see them as a certain type of human being, that could have lived many years ago and died out. There is much difference amongst the human species, as it is, and we know that there have been many groups of people who died out in previous times, so that is a possibility. In fact, how can one make up such a detailed account of the history of humanity based on some old bones. It has also been found that some evolutionists in their enthusiasm and desire for fame have placed the skull of a human with the bones of a monkey trying to pass it off as evidence – in order to be famous. Therefore, such claims to fossil evidence, are weak evidence, and not acceptable to build ones entire belief system, life, society, and afterlife on such assumption.

They claim that modern humans began around 600, 000 years ago, they say Homo Sapiens existed in Africa and Europe, this ancient species maybe became Neanderthal man in Europe then around, 400,000 years ago and anatomically modern humans in Africa around 200,000 years ago. There was something discovered by Richard Leakey, Kenyan paleo-anthologist in southern Ethiopia, they say around, 195,000 years ago, which they claim is accepted by many to be the earliest fossil of a modern human. They also say the fossil evidence as well as climatic evidence suggests that modern humans expanded out of Africa between 50,000 and 80,000 years ago , and that they spread out of the Indian ocean to Australia, and northwards into Europe, northeast Asia, and later into America. However, this really is all assumption, as how can one be certain from fossils, which may not be the oldest or demonstrate everything. In fact evolutionists themselves don't seem very certain themselves of such facts, and give facts out using terms such as maybe, possibly, could be... as if to show the element of doubt that it is not really certain evidence or proven, and it could really be something else.

4
HISTORY AND ORIGIN OF THE THEORY:

We shall see that throughout history, there have been different ideas surfacing about where we humans came from in our origin. We shall see that some of these stories involve animals and most are not really believable, although at the time may have been accepted by the majority of people. That is why one needs to look at the theory objectively. The intellectual and religious circles of Babylonia were, apparently, the first to initiate thinking about the human race. They brought forth the theory that one of their so-called gods' iea, created man out of the blood of the impregnated so-called god, Kalanga. They believe that man came into being when the anger of iea descended upon kalanya who was lead to the slaughter house with his sins and supposedly had his throat cut with a knife. Seven thousand years ago the Egyptians believed that they were the offspring of a cow. According to the second Egyptian story, man was born out of the ribs of a cow that emerged from the river Nile. The old Hindu civilisation belonging to eras of the rig veda and the yajur veda called the cow, the mother. The author of *the World of the Past* thinks that Greek philosophers were the first to propound a hypothesis about man, different from the one followed by idolatrous races.

There was also some mention about evolutionary theory among the ancient Greeks, which is interesting because apparently Darwin read the works of the ancient Greeks and so could have got some of his ideas from here. In seven hundred B.C. two pioneer philosophers, namely Anaximander and Archelaus declared that human beings had to pass through an evolution – the earlier generations bearing resemblance to a variety of animals. The first of the two claimed that man made his first appearance on the world stage in the form of a fish. The second thought that when the Icy crust enveloping the globe melted away and the earth, relieved of a massive weight, became warmer, life germinated on it and some living beings appeared on it of their own volition. Archelaus theorised that these living objects mixed sexually over a long period. In the last stage man branched out from the rest and formed a distinct identity. The author believes that these two could not convert many people to their views. For a long time no one gave thought to evolutionary theory. However, it was in 1843 when Charles Darwin and Alfred Wallace first propounded the theory; people had forgotten the previous work of the two field philosophers. Both were scholars of Greek and the study of Greek philosophy was their great occupation. Sir Arthur Keith says that a large number of scientists voluntarily dedicated themselves to following it and still now, after a passing of a century and a half after the original theory. One may wonder why they chose to accept such a theory. One argument is that the practice and adherence to religious ideas was declining at that time, and scientists since the enlightenment had begun to formulate their own ideas, which sometimes were contrary to the established church. Some of the things that led to this were the Copernicus theory that the earth was no longer central in the universe. So although many scientists continued to believe in God their concept of God was usually less personal and they did not hold God to be intervening in the world to such a great extent. Darwin himself was a Christian in the beginning. It was in this scientific environment, in which things in general were seen to be developing that such a theory could take hold. There were scientists in his time who were theorising that different species of plants evolved from one another and they were hence categorised into families, although no one went so far as to suggest that humans had evolved. There were other reasons also, such as lack of the technology that we have currently. In Darwin's time it was widely believed that maggots grew out of meat (rather than flies land on the meat and lay eggs) so general ignorance in scientific facts also contributed to the acceptance of the theory. It was on his journey as a naturalist on board HMS Beagle with Captain Fitzroy that he came up with the ideas for this theory. He apparently filled different notebooks with observations on biology and natural history. This material was not immediately seen as evidence for the theory. However, it is reported that later when Darwin came to survey and summarise his biological observations he had made on the voyage, it is said that he recalled three groups of facts, which, they say made it difficult for him to accept the immutability of species. The first 'fact,' for example, was in South America where he claims that he saw some resemblance between fossils of certain distinct armadillos and the skeletons of living species. Apparently the distinct forms were much larger. Later he claimed that these vertical successions were evidence of continuous descent with modification. The resemblance between historical successors, he claimed corresponded to the similarity between geographical neighbours. When he travelled across the South American pampas, he apparently noted certain forms of ostrich, which he claims were gradually replaced by distinct but nevertheless similar types. Each area, he claimed, was populated by its own distinct form. Later he did not see this as a result of separate creations but as the inevitable consequence of geographical separation. Migrating in different directions, he claimed, the so-called primitive ancestors of these two types had become so widely separated from one another that he claims they could no longer be freely interbred. The third so-called evidence that he provides was when he visited the Cape Verde islands off the coast of Africa and the Galapagos islands off South America. He thought that the argument from design would lead one to expect that all these island types would closely resemble one another, since they were created for the same physical conditions. He claims he was surprised to find that the similarities between the inhabitants of the Cape Verde islands and Galapagos Islands were similar because he believed they came from a comparatively recent common ancestor. Therefore, as one can see, Darwin had formed his theory by theorising. There has not been any evidence for the theory then, or now. However, his followers have taken it further than even he may not have done.

So what did the followers of Darwin do once they had accepted the theory for themselves? They travelled all over the world unmasking skulls, jaws and other body parts to attempt to 'prove' the theory. In their overzealous excitement they came to some conclusions about things that they had found, which they may not have otherwise have come to. When people are overexcited they may see what they hope to see, not what is really there. Such as people who see mirages of water in the desert, when thirsty, out of hope and not due to it being really there. They believed from this that the human race had to pass through four stages of evolution stretching over a period of one thousand years. Carleton, a British author described a century of scientific discoveries from 1858 to 1859 and divided man's evolution into four stages encompassing a vast period of 7,000,000 years. As Kausar Niazi (1975) mentions, to base conclusions on bones and Jaws, which are wrapped in mystery and about, which nobody knows who they belong to, carries little conviction. He asks why these scientists did not look at mummified monarchs and Queens of the Pharaohnic past and treasures of the royal cemetery before pronouncing man to be from apes and chimps. Are there no mummies among them to reveal what evolutionary stages man has passed through? Is there no evidence of the physical evolution of the body? As he points out science is supposed to belong to the entire human race; it is not to be put down to only one man's (Darwin's) findings. It was not Darwin or Wallace who managed to understand the secrets of nature. As he mentions, many people, many travellers, have come across skeletons and old bones, and none of them have come to Darwin's theory. He is right in saying that one is not authorised to create an imaginary 'race' and present it to the world without concrete evidence. He also goes further mentioning what historians have suggested of modern man's height, and that it has decreased in comparison with his ancestors. If you remember from the previous section, we mentioned that the Muslim belief is that people were bigger and lived longer than they do now. The four generations nearest to the first man, Adam – Ad, Samud, Qahtun and Jurham, he mentions, were much heavier built than the present generations. They were taller and lived longer. He gives some further evidence for this in the form of Yemenite caves. These were large caves built in the rocks and had high entrances, obviously corresponding with the height of the dwellers. Skeletons in the caves must have indicated that a much taller race must have lived there. Therefore, there is actually more evidence to imply that people of the past may have been taller and built larger dwellings for themselves than they do now, and that this is more strongly supported than the idea that the early humans were trying to learn to speak and get rid of their body hair.

Also when we hear of evolutionary theory we automatically think of Darwin, but he was not the first to propose a theory of evolution. Lamarck in 1800 had done so and Etienne Geoffroy Saint-Hilaire, another Zoologist at the National Museum in Paris in the 1830s. He apparently accepted evolution through his own research, where he compared the anatomy of different animals. The standard wisdom of the day was that animals are similar to one another when they function in similar ways. He thought that there were exceptions, such as Ostriches, which have the same wings as flying birds, although they do not fly. He argued that such things that separate one species from another were not entirely singular. He gave the example of a rhino whose horn; he argued was really just a clump of dense hair. The poet and scientist Goethe argued that the various parts of a plant – from its petals to its thorns – were all parts of one fundamental form – its leaf. European explorers were apparently discovering new species that Geoffrey thought fitted with his theory, such as the platypus in Australia, which was a mammal with a duck like bill, who laid eggs. He then held it to be a species between animals and reptiles – a transitionary form. There was also a claim that explorers in Brazil came across a lungfish that could breathe air through lungs, representing a supposed link between vertebrae's and in the ocean and on the land, instead of merely recognising it for what it was – a lung breathing fish. Scientists in England denounced Geoffroy as they had denounced Lamarck before him. One well known individual who attacked the ideas of evolutionary theory was Richard Owen in the 1830s; he was a young anatomist and the first to study new species, such as the lungfish and the platypus. These opportunities were used against the claims of Geoffroy. He was able to demonstrate that platypuses secrete milk like mammals and lungfish although they may have lungs, they did not appear to have nostrils, which all land vertebrae have. So he saw them as a type of ordinary fish. Although he abhorred Geoffroys claims he did recognise the similarities of some species. He believed, however, that Geoffroy had gone too far with his interpretation of evidence. He knew also that his research did not fit with new findings. Karl von Baer (A Prussian scientist) had found through his research that life did not follow a simple ladder, with more 'advanced' embryos recapitulating the development of what evolutionists term 'primitive' ones. Vertebrae's in the early stages were found to resemble one another, but only because they were just a handful of cells. They grew more distinctive as time passed. Fishes, birds, reptiles and mammals all had limbs and all initially formed buds as embryos. In time these buds would turn into limbs specific to the animal, such as fins, hands, hooves, wings and other sorts of limbs that may be unique to certain types of vertebrae. Therefore, although we may be made of a similar substance, it does not mean that we evolved from one another.

5

PROBLEMS WITH THE THEORY:

The main problems that I have with the theory of evolution, as you may have noted is both the fact, that the theory as a whole does not make coherent sense, neither does it explain reality as we know it, or give a better explanation to the argument that we were created by God, it implies inequalities amongst humanity and there is a lack of evidence to even imply that it did happen. Whenever, I have asked evolutionists for evidence they go off into some sort of imaginary theorising, admitting that there is a lack of fossil records, but bringing me books with carefully drawn sketches of a hairy ape man, bent over, and then other types of men who get gradually more upright and with loosing hair (are you sure it is not just their age) these men are given names, such as Neanderthal man and so forth. There is no evidence that these graduations existed. These are just sketches in a book that anyone can draw, they are not photos. We see humanity as equal, we do not see some hairy bent over people grunting and the rest upright, agile, intelligent. All the evidence that they believe is in their favour really has not proven anything other than that there is great variety amongst the creation of mankind. There are perhaps more types and animal varieties than has been assumed, but drawing conclusions from that about common ancestors and intermediary stages is just not scientific. There is no basis for that and it is wrong to mislead the public with what may be rather than what is. Evolutionist literature is full of 'maybe' and 'perhaps' and 'could have' statements. Evolutionists think that being designed does not require a designer (which is both illogical and quite absurd to be honest, like saying that this book is written and does not require a writer). They claim that there are lung-breathing fish that existed before air breathing land vertebrae's existed, and they argue there is still primitive, air breathing fish alive today. Well, does not that demonstrate that they are one variety of fish? I fail to see how there being breathing fish and land animals mean that land animals evolved from breathing fish, it is just not logical. What is the evidence that one precedes the other? Telling us that 360 million years ago 'one lineage of air breathing fish began spending some time on dry land (Carl Zimmer p.326)' is not right, because none of the present day scientists are able to really state what exactly happened back in that period of time. That sounds more like a story from Sharky and George or some other cartoon. Are they going to tell us next about the conversations these 'air breathing' fish had when they got tired of swimming and decided to dry out on the land? Perhaps they appreciated the fresh air?

In the fossil record there are conveniently no traces of these 'intermediary' forms that there should be if we really had evolved and if this Neanderthal man and other 'in between' stages actually existed. Yet there are evidence and skeletons of dinosaurs, which were supposed to have been living along time before all these changes and adaptations were alleged to have taken place. In 1856 a miner in Neander valley in Germany unearthed pieces of a skeleton that was then named 'Neanderthal man.' The argument was that since its jaw was massive and low, which they argued, raised the question over whether it was a separate species or one of extreme human variation. Personally I cannot see how one can truly tell for sure what something is, since it is a pile of bones. However, there is so much variety in the human race as it is, with so many different genes and traits that there is nothing so unusual about a skeleton with a larger jaw. Among humans there are tall and short in height, big or small in build, those with wide foreheads and those not so wide so a human with a wide jaw is not so much of amazement, even if it is not statistically normal. Neither does such a finding indicate that this is an intermediary form or some half evolved man. There needs to be a lot greater evidence than mere guesswork on the part of such discoverers. Finding one skeleton, that one is not entirely sure about, can hardly be evidence to base an entire theory. The other things that explorers uncovered were fossils and tools, which showed nothing of evolution. Therefore, Darwin and his followers have only been able to compare humans with apes. His argument was that bone for bone, humans and apes are identical. Whereas this might be the case, we cannot then conclude as Darwin had done that this means we evolved, because it does not. I could equally argue that

God created us, physically, with a similar bone structure. This would actually make more sense. Also, the bone structure is not identical. Apes have long arms and short legs. They hop around using their arms. They are bent rather than straight and are designed for living primarily in the trees. Humans on the other hand have long legs and shorter arms. We tend to walk up straight (well, most of us anyway), so we can expect that our structure will reflect this difference – we are not designed for living in trees (I have tried many times as a teenager and hurt myself in the process), but for walking on land. Evolutionists further claim that from studying genes, the supposedly most recent 'ancestor' of humans and chimps lived about 5 million years ago. Therefore there should be more evidence than there is. The other issue is that if these 'common ancestors' died out (and conveniently they all died out, leaving no trace anywhere, even in the fossil record that they ever actually existed) then how come the chimps that we see are still here, swinging away in their trees? Should they not have died out as well? Since according to Darwin, things evolve to suit the current climate and needs of the creature. If he claims that humans had to evolve because it suited us to do so, then these 'lower life forms' should, according to his theory, have become extinct (or evolved along with everything else) and actually the so called 'intermediary forms' should have stronger abilities than chimps (they should have still attained their abilities in the trees and had begun to walk) so therefore, they have more reason to be still around than the chimps, yet conveniently there is no sign that they ever were there in the first place. There are other claims that fossils have been unearthed, such as in Ethiopia, where a fossil was uncovered that we are told looks like a chimp but seems human. Other scientists found fossils dated several millions of years in East Africa. These all present the same problem – that of interpreting the evidence. I mean, if there was no Darwin, and no one came up with the idea of evolution, would people really be confused and discussing whether these fossils are intermediary forms or not? It is unfortunate that people who go on expeditions already believing in evolutionary theory are going to be looking for evidence. When one is looking for such evidence for ones beliefs it is probably easy to see such skeletons as supportive. Those of us who do not go out looking, holding that belief cannot see such things as evidence; we need a lot more than your understanding of findings.

The other problem that relates to the actual lack of evidence is that when there are things that could possibly be taken as a proof, there is inevitably a person, an understanding and interpretation behind such beliefs. Therefore, some of the claims for evidence are actually due to one's own understanding or interpretation of what something implies. Let us look at some of the claims of evolutionists and how things could be understood differently. They claim that findings of palaeontologists of whales with feet are evidence for evolution and that they must be ancient cousins of today's whales rather than direct ancestors. It is obvious actually those whales must be a particular type of whale rather than a whale in the process of evolving. It is perfectly reasonable to argue that God may have created some whales with feet and some without. The 'feet' that they say they have could be there to help propel them through the water or for some other benefit. They may all be relatives without sharing a common ancestor. Just arguing that there is an evolutionary tree and saying how whales moved from land to sea does not mean that they did so, because no one was actually there to see this happen, so it is purely imaginary musings and not a scientific premise. It also does not explain why they would do that or how they would develop these supposed abilities. I mean, if I want to live under the sea, how am I going to develop the ability to breath under water? And what about developing some fins? Can I choose to do that or is there something or someone who will choose for me, or if I lie down in the water and wait shall I develop this ability, but then I would have drowned! It is claimed that whales are related to cows and hippos. I cannot see the likeness. Carl Zimmer states that 'the transitionary fossils of whales show how the transformation could have occurred in small steps (p.325)' everyone knows that there are no transitional forms in the fossil record; even most scientists will admit that. Also, Michael Behe, another scientist, but one who does not support evolutionary theory argues that it is not actually possible for things to have come about with small changes and so forth. There would have to be giant leaps. When one puts a bike together for example, the parts are needed whole, if they are tidy bits of parts of the bike they will not serve any purpose and will be of no use. This is why a bike has to be put together with wheels and all the parts as a whole for it to function as it should. Carl himself had earlier stated that there is a scarcity of fossils. He states that 'Palaeontologists have found many intermediary forms that creationists claimed could not exist (p.325)' however, what makes him, or anyone else so sure that such 'findings' are intermediary forms? How do you know that such skeletons are not just another species or a different breed of the same thing? God has created many differences amongst human beings and also amongst many species in the animal world, therefore, is it not more likely that whales with feet are what they are rather than concocting imaginary dramas about cows and hippos walking around, gradually changing to blue colour, getting bigger and transforming from a land creature into a water creature, losing everything else but the feet. Does the man have any idea how absurd that sounds? It makes no sense at all. They claim that the origin of swimming whales seems to have taken place over millions of years. Evolutionists love to say this in order to put difficulty into attempts to disprove it. One may well ask not only how this happened, but why? I mean, forget for a minute how this cow became a whale, but what would be the reason and advantage for that to happen and then if there was such an advantage, why have not the others developed such favourable abilities. Cows seem to be quite happy to me. I mean in all the millions of years we can document cows have given birth to calves. It is automatic in English that the child of a cow is a calf.

It is not a whale or anything in between and we have never witnessed anything different in calves that are born because they inherit genes from their parents. They do not just evolve. There are other claims by evolutionists as to why changes happen. They argue that as the climate cooled, the bodies of ancestors changed. They claim that toes became like fingers and they grew longer legs, they argue that they held their back and head upright. However, have we ever seen such changes taking place? I mean, as I have already mentioned, as far back as we can trace humans have been entirely human, not anything in between, so we can only classify this story as a type of myth or science fiction. Keith Hunt from Indiana University puts these so-called changes down to diet. He claims that the first hominids were probably on all fours. The biggest change, according to them was walking upright. They imagine (because that's all that one can do without evidence) that they were forced to walk more slowly, and so covered only a short distance each day. They imagine these strange creatures slowly moving from one tree to another picking the lowest fruit. One wonders how ones imaginary musings have been accepted by scientists as actually happening and how this has come to be accepted by the British public as actual history and fact. It is scary really. I mean, imagination is meant for art school, not science. They even use the language themselves, they *imagine*. We cannot know for sure what really did happen, since we were not there, but since there is no other evidence to back up their imaginary scenario we will have to reject that. We have never seen it happen, or any evidence for it and it remains improbable to the sound mind how it could happen, so my conclusion is that it did not happen. The language they use can easily enough be changed to support design and creation arguments. When they say such and such evolved that way, we can very easily say that such and such was designed that way or was created that way, which actually makes more sense. The argument is that everything takes thousands of years. But then it cannot benefit the species when they need it. Neither I nor my offspring will benefit from supposed 'evolution' if the climate changes now. And probably in another thousand years, if people are still here, there will be more changes in the earth and atmosphere, so tiny changes from things now will just not benefit. We see children born generation after generation without anything different about them – they are clearly human babies.

One of the main beliefs of evolutionary theory is that of natural selection, which I would also like to argue against. The idea of natural selection, that competition between two species can drive one to extinction does not really prove evolution, because from what we see around us, we see that for example, Cod is in danger of becoming extinct because we eat it too much and other animals because they are hunted. This does not mean that one species is evolving more than another; rather it is people's greed and eating a lot of this fish that is making it less. In fact, in both the animal and human world, there is a mercy and understanding amongst one another that Darwin failed to recognise. Most human beings will help and support a human being who appears weaker and more vulnerable than themselves and even animals in the same category. That is why we have charities to help the destitute, the orphans, the disabled, the learning disabled, the mentally ill and the old people. We even have organisations and societies to look after the welfare of animals. Many of us help animals when we find them sick, although they are not our species and we know they will never return the favour. Rather such an act is one of charity, altruism and selflessness. Most of us do not want to be in competition with one another fighting for survival. We have wars, but these are usually ideologically based or because people want land actually rather than for so called 'evolutionary' reasons. One of the final reasons for my rejection and disbelief in evolutionary theory is because it contains in it a tendency to racism. When you believe that everything is evolving, there will have to be some more 'evolved' than others; usually these have been considered by the Europeans and the Nazis used such a theory to support their campaign. Whether people who believe it are racist or not, it has that tendency and also because it encourages fighting and 'struggling' because it emphasises the 'survival of the fittest' so it could easily be used to justify all sorts of wars, fights and oppression, with oppressors claiming to be the 'fittest' fighting for survival. Therefore, what one believes can have serious impacts on an individual's mentality and on society; this is why I do not like it being propagated.

De-evolution from monkeys and a return to humanity:

6
WHAT THE THEORY MEANS FOR PEOPLE AND SOCIETY:

It is clear that the beliefs and ideology people hold will have an effect on one and ones relationship with others. Ever since the publication of the Origin of the Species, people have been considering what evolution really means for them in their lives. Darwinism rendered mankind nothing more than an animal, so it denies man as a special creation. This brings morality and also religious belief into question. In Britain and the United States evolution was being used to justify laissez-faire capitalism. By 1870s almost all scientists in Britain accepted evolutionary theory, although they do not agree on how they believe that it unfolded. Most Americans accepted evolution by the end of 1800. The British Philosopher Herbert Spencer promoted a mixture of Darwinian evolution and Lamarckian evolution, and claimed that free market struggle would make humans evolve greater intelligence. He acknowledged that there would have to be suffering along the way - Spencer thought that Irish famines were an example of people taking the high road to extinction but he held that this suffering would be worthwhile because humanity would be elevated to moral perfection. He won admirers all over the globe. His admirers coded his ideas into a school known as social Darwinism. This school held that at the distinction between the rich and the poor at the end of the 1800s was not an injustice, but biological. Social Darwinism took natural selection away from biology and into the social environment. It lent authority to government efforts to control the supposed 'evolution' of the human race. In the early 1900s retarded people were sterilised and others who were deemed to be degenerates so that they could not contaminate the evolution of their nations. Statesman William Jennings Bryan was hostile to evolutionary theory as he says "because I fear we shall lose consciousness of God's presence in our daily life, if we must accept the theory that through the ages no spiritual force has touched the life of man and shaped the destiny of nations. But there is another objection. The Darwinian theory represents man as reaching his present perfection by the operation of the law of hate – the merciless law by which the strong crowd out and kill off the weak (p.318)." Carl Zimmer argues that he would have been correct if he had been referring to social Darwinism or monism which people used to justify brutality, poverty and racism, but he argues that both philosophies were based on misreading of the Origin of the Species, mixed with Lamarckian pseudoscience. He claims that Bryon did not see this difference. To be quite honest I myself do not see much of a difference. I mean I mentioned how the theory influenced my perceptions of other nationalities and others as well. In school I was only taught about the biology of our supposed origins. Our biology, however, is going to affect other aspects of ourselves as well as our way of seeing the world. To me all the evidence points towards us are being created in our entirety by God. I cannot see any evidence that we did evolve. Monkeys in the zoo are only monkeys in the zoo as far as I am concerned. As far back as we can trace our ancestors were not found to be hairy apes swinging in the trees, so why should I believe that? And what has happened now did this process of so called evolving suddenly stop? And did those who evolve first wait for the rest to catch up or are some of us superior to others? It is very hard to argue that humanity is equal when we believe we evolved. When we came from two human beings we can believe that humanity was equal and we were born and created in the same manner.

7
WHAT THE THEORY IMPLIES FOR HUMANITY:

What the theory implies and means for humanity:

In terms of what belief in evolutionary theory means and implies for humanity, it implies inequalities and some element of aggression and selfishness. If people believe in survival of the fittest, they believe that to survive they have to fight and overcome others to be strong. Their belief in evolution means that some people will be perceived as less evolved and others more evolved than others, which often results in elements of so called racism (there are no 'races' in humanity). Usually this is belief that whites or Europeans are superior to black people or Africans, which they perceive as more similar to apes. Evolution played a role in disturbing cultural changes in both the United States and Europe. During World War I it was used by the Germans to justify the carnage carried out by them. They followed the German biologist Ernst Haeckel who held humans to be the most superior of all creation and claimed that we are still developing our mental capacity. He saw some humans as more progressive than others and divided them into 12 different species ranking them from lowest to highest. The Africans and New Guineans were held to be the lowest and at the very top were held to be the Europeans – "homo mediterraneous" within the Europeans; he placed the Germans at the very top. He stated "it is the Germanic race in North-western Europe and in North America, which is above all others, is in the present age spreading the network of its civilisation across the whole globe, and laying the foundations for a new era of higher mental culture" (Carl Zimmer, 2001 p. 316). He held that eventually all other 'races' (as he called them) would "completely succumb in the struggle for existence to the superiority of the Mediterranean races" (Carl Zimmer, 2001 p. 316). As we have mentioned, evolutionary theory does contain in it such a tendency towards racism. I am ashamed to admit that while I held that belief that we came from monkey men, I could not help but think at times that Europeans were more evolved than Africans and others. Obviously the media and the European mentality contributed to that (such as showing the British as others developing technology and being superior to others, while others were shown as underdeveloped and primitive). I am being blunt here (and I really hope that no one takes offence when I say this, because this is NOT a belief that I hold now) that I thought that Africans and other black people were more similar to apes than we were because, I thought at the time, I had not seen a white monkey, they were brown in colour, therefore, I thought, the skin colour of brown people showed them to be more similar to these apes amongst other things. When I once said what I was thinking out loud to an older woman at the time (I believe I was about sixteen when I was thinking about this, as I had been hearing about evolution more in college) she agreed with me but told me not to say so in front of black people because they would accept it or like it. I feel bad about holding such a belief now, but such a conclusion is drawn from its premise. In fact, a Somali friend of mine once asked me what I thought about evolutionary theory and what it made people think about each other. When I told her that I do not believe it in it and it does contribute to racism, she said she had thought so and told me a story about an older woman (she says she was in her fifties) who said about her and some other Somalians that some of the problems that they had (and I cannot remember exactly what these problems were but they were related to being late, or something similar) were because of evolving that way, that they were not as evolved as white Europeans. My friend mentioned that she felt hurt by what she said, but she was told by another person to ignore her because she was older and because she believes this theory. I know of other African people who do not like evolutionary theory because of what it means – and peoples understanding in relation to them. Perhaps not all people who believe in evolution are racist (at least not openly) but the theory itself certainly does take one in that direction.

Physically we are more varied in appearance than apes, having blue eyes, red hair, blonde hair, brown hair, black hair, brown eyes, green eyes, grey eyes, black skin, brown skin, and different shades of paler skin or darker skin. We have eyebrows and they just have a protruding shape above their eyes with no hair on their faces, they have a different more rounded and carved look to their faces, and they all look very similar, they all are brown colour, not like us. Are you telling me that this was merely the result of random mutations? Their life is swinging in the trees, eating and socialising. We human beings have set up schools, we have (well they claimed we have) gone into space, we have built aircrafts, cars, computers, we have methods and machinery for farming, we have a language, we do not have hair on our bodies and so make and manufacture clothes. We have religion and we wonder and contemplate about our existence and how and why we got here. Therefore, if you look at apes and humans – in fact all other animals and humans one will see that we are very different. Physically and in other ways; having so many digits similar does not mean anything. Technology separates us – we have created computers, as I am using now, cars, phones, televisions, space ships, ocean ships and so forth. We are not born with hair on our bodies; we must make and form our clothes ourselves, making different styles and varieties, choosing different colours and materials and competing in fashion parades with them. We do not (usually anyway) merely eat foods from the trees, on the ground or on bushes, but we concoct different varieties together, decorate food with herbs and spices, cook, fry, bake or boil it – animals do not do this. We have a desire to learn and know about the world. We learn about the world and transmit this knowledge and information in both the spoken and written word – animals cannot do this. Our desire to know has taken us up into space after the creation of the space ship and into the depths of the ocean. We transmit what we may find to others because we have a strong capacity for language and communication, which is another major difference. I remember one of my psychology tutors who was discussing language ability in humans saying that only human beings have this capacity for language – no animal has language. They may squeak and grunt to indicate something (which their fellows may understand) but they do not have the ability to hold conversations and learn the vocabulary, as we know it, sometimes learning several languages. The capacity to relate and use language demonstrates another way in which we are different; we are actually social beings more than animals. Although animals have herds and groups and packs, they do not have the tribes, nations, alliances, social groups, support groups and clubs. They do not separate themselves off as we do. We have these groups and they tend to separate one from another. Another way we are different is our capacity for will – although animals clearly have a will, otherwise they would not be able to do anything, it is not likes ours and not to the same extent. Throughout our lives we are forced to make choices how to live, we fight for causes whether just or unjust and are able to carry out actions whether good or bad. Actually the ability to distinguish between good and evil and then act accordingly is another ability that we human beings have. We humans carry out the most crime – murders, rapes, stealing, lying and other injustices just do not take place in the animal world, at least not to the same extent (humans can be more barbaric and have made guns and bombs enabling them to cause even more deaths and damage) and with the exception of the law of the jungle. They also do have courts of justice whereby members and fellows of the species who harm others or the social order are tried and punished. They do not have laws and morality. We humans also have a greater capacity to do good and look after one another and even other species. We fight on behalf of animals that are being unjustly killed in other parts of the world and hold campaigns to protect them against extinction. We keep animals as pets and look after them. We fight for the welfare of other people's children, not only our own, for the elderly, the disabled and so forth. We give in charity, erect buildings to look after the sick, the orphaned, the homeless, and hold animal centres to take care of animals in need of care. We do not expect animals to repay us for this. Finally, our morality, and actually our religious and spiritual capacity sets us as different to apes and swine (well, most of us anyway). Animals do not search for their origins, worry about where they will go after death, pray and follow rules and rituals of a particular religious denomination. There are no ape like priests, rabbis and imams. They do not fast or single out specific days and times for religious worship and duties. There is no argument about where we came from and where we will go and there are no politics, elections and arguments about who is right and who we should follow and so forth. Animals follow the laws God put in nature without argument, it is only we who argue and make a fuss. None the less, despite all of these very obvious differences evolutionists hold that we are just overdeveloped apes and any difference that you see is due to evolution. The main problem may be that they focus on physical differences and have ignored the spiritual, mental and emotional differences of human beings and our great capacity for behaviour. Evolutionists claim that evolution shapes our psychology affecting our capacity for love, jealousy and other emotions. They believe that such capacities are merely inherited from our ancestors.

Evolutionists have also referred to some Africans as 'hunter gatherers' they say:

"Even today, hunter gatherers with much more sophisticated weapons like poison tipped arrows do not catch enough game to feed their families. Kristen Hawkes, an anthropologist from the University of Utah has studied the diet of Hadza, a group that lives on the East African savannas. While they occasionally eat a gazelle or some other big animal, they depend on roots and tubers for a steady supply of calories." The implication here is that in parts of the world where people may live a more natural lifestyle, simpler and in tune with the natural world, the less evolved they are. I wonder, does such an idea also come about because these people were African and most likely black? The next page informs us that I million years ago these 'hominids' as they call them, began moving into Asia and Europe. In the beginning we are told that they did not go further than the south of England, and then, after another thousands of years they began to go north. Again we have to ask how these evolutionists know so much about who went where exactly, how they were and how long it took, since one thousand years ago would not leave much of a trace. It also demonstrates that the whole story of evolution gives an image of the 'developed' 'civilised' and 'superior' Europeans and the 'primitive,' 'backward' and 'less advanced' 'hunter gatherers' who still have some catching up to do and who remain 'less of a human' to put it bluntly. Is it just me, or does the theory seems to breed a racist and generally discriminatory interpretation of facts and reality? It is not only myself who understands from the theory these ideas.

8
HOW EVOLUTIONARY THEORY HAS BEEN INFLUENCING ALL AREAS OF LIFE:

The Times: Monday February 23rd 2009-05-11

Article entitled: It's not teenage strop. It's a key part of evolution (they say).

David Bainbridge says that there are good reasons for teenage moods, which, according to him, are caused by developments in the brain vital to man's success as a species. the article is full of opinions, conjecture and stereotypes. He claims that teenagers are "self-centred, stay up to late, sleep in, shirk their responsibilities and spend too much time with equally antisocial friends." Now, has he met all teenagers to be able to come out with such a statement? Does he know all the teenagers around and their routines? Did he not meet the good and well behaved teenagers who spend their time studying, working or in different types of voluntary work? And perhaps the very expectation placed on these young adults just might be an influence on them. In fact the term 'teenager' itself, is a modern term, that is made up and has no real meaning by itself. He further claims that adolescence has not been understood very well until recently. However, he argues, new studies in psychology, philosophy, neuroscience and palaeontology, he claims, have revolutionised our understanding in the past ten years. He then jumps right in with the assumption – an unscientific one – saying that "before that we did not know when teenagers first evolved." Now, from where comes from his certainty in the beginning, that they did actually evolve. It is also a rather strange statement to make, as if implying that teenagers are somehow not part of humanity. Perhaps because he is a vet with a zoology degree, he is interested in seeing similarities with animals, but teenagers are not animals in the zoo, nor are they swinging through the trees and there does not seem to be any more or less of a development that any other type of human being. He claims that teenagers behave the way they do because of changes in their brains. Now, firstly, 'the way they do' is according to his stereotype of how he thinks that they do, or should behave – all of them together, secondly, how can he be so sure that all teenagers all over the world are the same and it is due to their brain. What about culture, diet, the stereotypes thrust on them, unreasonable demands and the confusion put onto them that on the one hand they are young adults able to make their own decisions and life choices and on the other that they are still children who must obey and be treated still as children. He further claims that evolutionary theory suggests that the teenage years are the best years of our lives, because they were when we used to select mates, he says, and start having babies. This is ironic considering the fact that the government holds it to be illegal for teenagers to have relationships and babies under sixteen. However, we quite clearly can see that they are entirely capable of such things. What would be wrong, however, in saying that God created us with certain desires and needs and especially at certain times, rather than saying that these evolved. Would it not make more sense? And anyway, one has to ask how our basic desires and needs were part of a slow and gradual evolution – how would the 'species' as he calls us humans (must be the zoology) continue itself in the meantime, if not able to pass on offspring, do they say that animals partly had children with what they had? And were these children incomplete in themselves? All of these types of detailed questions need answers if we are going to be continually bombarded with evolutionist jargon claiming it to be a fact influencing so much of our lives. Another claim he makes is when he states that *"Recent discoveries in palaeontology allow us to measure how rapidly our hominid ancestors (he claims another assumption) grew up. As humans mature, tooth enamel is laid down in a regular daily cycle that leaves an imprint in the layered structure of the tooth, so we can "age" fossil teeth by sawing through them and counting the rings, in a similar way to determining the age of trees. The results suggest that the first "numerical teenagers" – the first humans to take that extra decade to become mature – appeared 300, 000 – 500, 000 years ago. Tantalisingly, this was just before the human brain made the great leap to its full Homo sapiens size."*

Now, it's hard to see how recent discoveries in palaeontology could find all of that! The idea of having a hominid ancestor has never been proven, it is also unlikely that if there ever where such an 'ancestor' that we would be able to know how it grew and how long it took to grow as well. Finding how teeth aged hardly can determine that human beings evolved the ability to mature! In fact, at certain times in history life expectancy was shorter due to food shortages, therefore that is one explanation for the findings of the teeth, another is that it may not have been a human tooth.

The problem is that evolutionary theory has been influencing so many different theories and subject areas that it has meant firstly that many previously spiritually orientated disciplines have become materialised and secondly that beliefs and assumptions are based on these supposed monkey men, who we would argue do not exist (since there is no firm clear evidence of their existence), which makes the whole foundation of any arguments meaningless and futile. It has the further effect of not allowing any spiritual interpretation and so ostracising a great amount of people, from certain subject areas, such as medicine. To give an example, Patrick Holford, in his optimum nutrition Bible, states in an opening chapter entitled *From Monkeys to Man – Nutrition and Evolution* that "You are much older than you think. The human body you walk around in is at best the result of millions of years of evolution (assumption), the vast majority of, which was spent living in an ordinary environment and eating a diet very different from those of today. Understanding the dynamics of our evolution can provide essential clues for promoting health." This is his opinion, but what it means is that basing studies of health on an unproven theory could likely lead to many other incorrect assumptions and interpretations. If the underlying assumption about what human beings are and where they come from is incorrect, the assumptions about health or other things built upon would most likely be incorrect. We would also be able to say, without all the evolutionist jargon that in the past people lived in an ordinary environment and ate food differently from that of today without referring to evolutionist theory. We could even say that they led a life more as God had created for them. Evolution really does not explain anything more, or add anything to this sentence.

A final point I would like to make about the materialistic outlook and the effects it has on science, psychology and general treatment of people, is that when one fails to acknowledge the soul, indeed, the essence and humanness of a person, it effects the way the person is treated and perceived. If they are sick, then one may ask what affect this may have on their illness. Larry Dossey mentions in his book *Healing Beyond the Body* that even students in training become doctors and doctors themselves are affected by this outlook, and he mentions how it disheartens many and leads them to loose motivation and extinguishes much of their desire for healing and caring for people. He mentions that the concept of healing has gone from much of contemporary medicine, and much of the interest in alternative medicine comes from a desire to bring back some meaning to medicine, since it has been removed from medicine and society at large.

In terms of psychology, it is my belief that the medical profession could do with bringing back the soul of the person to their body and taking it into account. Instead of acknowledging the spiritual aspects of a person and perceiving their emotional and other problems as possibly spiritual based, they deny this and categorise people into 'cases' and diagnose them with labels that do not necessarily help in bringing about a cure.

Scientific assumptions occur again when we hear that someone who is a schizophrenic is genetically so (and usually 'genetic' is used to mean someone is just 'born that way' and is beyond help), and when they are told this, it means that they are often on drugs for life. Having worked with mentally ill service users for the past few years I can say that they usually do not complain about the various people helping them. Doctors, nurses and social workers, they have told me, are quite caring and supportive. However, from what I have seen from the ways certain individuals speak to these patients is that they lead them to believe that they will not get better, that they just have to deal with it. I heard one CPN say to a lady, when she asked her when she would be better that "you may never get better" perhaps she did not want to give the woman false hopes, but still, she could have phrased it in a more positive manner. I thought that the whole idea of doctors and nurses, in fact the NHS, was to help people to get better, not to drug them until death. I also thought it insensitive to call her adopted sons adopted mother 'mother' in front of her like that, since she is the mother. Such people are weak and cannot stand up for themselves so much, so it is very shameful that they are being oppressed in this manner. The fact that these individuals are usually given medication that often has horrible side effects, and that some of them told me do not take away the problem anyway is another factor I deem to be inhumane. One lady told me her medication makes her dribble, and she has developed diabetes as a result of anti-psychotic drugs. Another told me that hers makes her sleep very heavily so she can hardly move when she gets up. A man told me something similar, his medication made him feel heavy so that he could not get out of bed, even his walking was slow. There are many negative and horrible side effects of these drugs, which from what I have seen do not help, if anything they only add to the suffering. Why not try more kind treatments, talking therapies and spending time with the person instead of confining them to a life of medication, which I do not think is all the time needed. Perhaps in serious cases, when no other method could be found, and when the person is at risk of harming themselves or others, then medication could be used as a last resort, but it still does not get to the root of the problem, neither does it cure it.

In current psychiatry, however, there has been an increasing tendency to link various creative impulses with madness and to pathological brain chemistry, neuroendocrine disorders and genetic factors. Psychiatrist Arnold Ludwig of the University of Kentucky Medical Centre in Lexington found that one third of the eminent poets, musical performers, and authors of fiction suffered from psychological problems as teenagers. This rate was three fourths in adulthood. Prominent scientists had less mental disease than artists but there was still a steep rise in suicide rates in old age. All experience researchers have now put it down to brain chemistry. For example, the feeling that creative individuals get that they have tapped into a higher, transcendent reality is being connected with chemicals in the brain. Researchers M. A. Persinger and C. M. Cook of Laurentian University call these feelings "sensed presence" – basically a sense of the Almighty. They claim that this feeling originates in the right side of the brain and can be produced artificially. They exposed subjects in a laboratory to electromagnetic fields and asked them to press a button when they felt "a mystical presence." The subjects did not know when the exposure occurred. More often than chance would predict, mystical presences (button pushes) correlated with the application of magnetic fields. These results led some people to believe that religious experience is brained based. Actually this study could not really test for mystical experiences because is it possible for someone to choose when they have such an experience. Also, the participants may have been responding to the electric shock because that was the only thing that they felt, perhaps they did not have any mystical experiences, so there is no similarity there. Such experiences can also not be said to be purely brain based because the brain responds to external stimuli. When we see something the brain responds and things happen in the brain. It would not make sense to say that the brain started the movement and made up what we are seeing, so it is actually within the brain. Otherwise we may as well argue that we are living in a big brain and that when we die or when we sleep the world is no longer there. But this does not make any sense and is very egocentric. Neurologists have gone and linked hyper religiosity - heightened interest in religion, with temporal lobe epilepsy. Well now, that just leads all the priests, rabbis, imams and saints towards diagnoses, how convenient for the secular minded! Actually surveys show that the majority of Americans believe in God and have a "sensed presence" of a non-local, infinite Almighty. Do most of them have temporal lobe epilepsy? Scientists have been writing off religious states to abnormal processes in the brain for over a hundred years. Larry Dossey adds that if we take these 'scientific' findings seriously then it might appear as if our creative impulses were rooted solely in the brain. However, none of the research mentioned is incompatible with non-local factors in creativity, such as sharing thoughts with distant minds, participating in various mental fields, dipping into a universal cosmic soup of intelligence, obtaining insights from a higher source of wisdom, and so on. If we accept that creativity is purely brain based we give up too much evidence favouring a non-local side to the mind. No one in the entire history of science has ever proved how the brain could produce a thought, creative or otherwise. The brain is only matter, and impulses. The brain is a transmitter, not originator of consciousness. Whereas the brain may affect the contents of consciousness, there is no evidence at all that it produces it. Nancy Murphy, philosopher of religion at the Fuller Seminary in Pasenda, California says, "If we recognise that the brain does all the things that we [traditionally] attributed to the soul, then God must have some way of interacting with human brains." She says that neurological research can be seen as an attempt to give "an account of divine action.... -how God acts on the brain."

Elio Frattoroli argues that Psychiatry is supposed to mean, "healing of the soul." This is a call and need that is not fulfilled simply be prescribing medication. Elio believes that healing the soul requires a growth enhancing personal encounter with another human being in a psychotherapeutic process. In what Elio calls 'the age of the brain' (our current time) patients are objectified, diagnosed as "cases," equated with their brains (and genes), and treated according to a standard of statistical science rather than personal knowledge. The doctrine of scientific materialism – the assumption that mind and soul are products of brain activity – makes it easy to persuade ourselves that there is no mind body problem, no inner conflict or self-alienation and no need to examine the inner life of a soul. He believes that this philosophy is destructive to western culture. By denying the dualism of our inner experience, it gives way to a tendency to denial – not only of inner conflict but of inner life all together. This has led to a politically endorsed, corporately sponsored psychiatric drug culture. He argues that if we were going to look, without the influence of materialism at the current craze for psychiatric drugs, we could see it easily enough as a symptom of a sick culture seeking to anesthetise itself from the pain of human existence any way it can – through cosmetic pharmacology or cosmetic surgery, alcohol, drugs, gambling, promiscuity or violence, through mindless television or mindless cults. Each choice makes us feel emptier and uneasy. These represent the mind body problem – the impulses of the flesh against the needs of the spirit. Materialists tell us that there is no spirit, which makes it easier for them to deny the sickness of the soul. We are still left with a persistent aching of the soul and its unmet needs. Perhaps this is why I was warned by a tutor of mine in university, when studying for a degree to be careful about considering going into psychology when I have a religion, religious beliefs and outlook on life and I do not believe in evolutionary theory. He assured me that in the psychology department the majority of the people there were atheists who strongly supported evolutionary theory and he, as a devout Christian felt like an outcast. He informed me that he felt different both to the other Christians who were part of his Church and also to the psychologists who did not share his belief in God and the Bible. He had made some attempt to bring religious elements into clinical practice, but it did not appear that there was a great amount of success. Perhaps we need more religious people in such practices, rather than allowing them to be scared away at the prospect of having to face materialist assertions and claims. There does need to be some defence and retaliation, does there not? And what about the people at the receiving end, the clients? If they would benefit from such religious values and practices being brought into place in psychological therapies then they should be encouraged and welcomed, not really allowed.

It should be clear that all elements of a person should be considered when such a one is suffering from any ailment, not merely the physical self. Perhaps it does not lie in brain chemistry, because any disorder of brain chemistry will most likely have an external cause anyway, such as stress and so forth. Thinking processes may not cause it, but negativity may be part of the symptoms. Rather than looking in the physical arena of a person for an answer (and not actually finding one) perhaps one needs to look towards the spiritual soul of the person. Perhaps it is actually the spirit that is suffering, and not the body. May I add that in a quest for truth and for the sake of making the patient better and not merely arguing ideologically about who is right or not, perhaps we need to re-consider exorcism and possession. I am not talking about beatings or harming the person in any way or anything seen on the exorcist, but since we must consider everything in a search for a cure, and there is nothing to say that it does not happen, since modern medicine is at a loss to explain much of mental health problems and hallucinations and delusions anyway (and many people and places believe that it does occur) therefore, it should be considered by health professions in an attempt to support the patient to better health.

9
CRITICISMS::

Perhaps a scientist will ponder and think about a certain issue, whereas it may not be as is assumed. Another problem can be interpreting data, for what does something mean may differ from one person to another. It may also be influenced by personal desires, beliefs and inclinations. Is it ever possible to be entirely objective? While atheists claim that creationists and believers in divine design are prejudiced and biased in their views, is not an atheist just as biased when faced with evidence eluding to divine providence and denying it because, to them it is unscientific? Is that not bias at its worst? What is more than that is when theories are put forward as proven and fact, when they are not so, and perhaps lead the masses astray. Is it really ethical to lead the general public to believe their origin is other than what it is? To teach the fact that has not ever been proven, and never will be and not give balanced evidence against it that disproves. All this in the name of defending atheism, which actually does not explain anything much or give anything much to people in the way of hope or happiness or increasing their quality of life. Where we hear from some scientists that they claim our origin to be from aliens we hear from others that it is from monkeys and animals. Both stories to the writer are absurd explanations of reality, that really don't explain anything much and leave more questions than answers. The alien theory is probably not so well accepted, however the theory of monkeys has been taught to us in schools as factual and true and stories of creation to be taught as myth. A 2007 article tells us "shock claim: Human DNA 'was designed by aliens' who spent 13 years working on the human genome project, (Sean Martin). In the article they claim aliens wanted to preserve DNA or simply plant life on other planets. a pair of scientists from Kazakhstan believe that humanity was formed by a higher power, which they believe was an alien civilisation who encoded symbolic language into our DNA. They believe 97% of non-coding sequences in DNA is a genetic code from aliens from which they claim that the mathematic code in DNA cannot explain evolution, it's a sign of something on a higher level. They say a message could have been planted by aliens. Furthermore, there are some who now explain some mysteries such as dark matter rays. They say aliens are everywhere (strange we don't see them then). We mention such things here, even if far out to demonstrate that such claims come from members of the scientific community in an attempt to explain reality and life. However, it is probably not so widely accepted as evolutionary theory. Therefore I hope to discuss the theory of evolution here as I have understood it and what has led to reject it in the end, hoping that this will be of benefit to readers.

Scientists are now currently divided into two sectors, or schools of thought or belief, although evolutionists may claim that all scientists accept and agree on the truth of evolutionary theory it is the case actually that many modern scientists are now disregarding the theory of evolution and atheism in favour of what they call intelligent design, and that is in reality a denial of evolutionary theory and confirmation of a Creator based on evidence. Since it is based on evidence and observation, it is held to be scientific, based on a four step approach involving observations, hypotheses, experiments, and conclusion. If everything is ordered and appears to be designed there must be an intelligent designer. Some of the more known scientists of this stance are Stephen Meyer who's book Darwin's doubt was number seven on the New York Times best seller list and provided evidence for the scientific theory of intelligent design. And in 1996 Darwin's Black Box by Micheal J.Behe of Lehigh university, argued against evolution on the micro biological level, and in favour of 'intelligent design,'. Adam Sedgwick, wrote to Darwin in his time that parts of it were completely false and filled him with sorrow. Astronomer Sir John Herschel rejected the theory calling it "the law of higgaldy piggeldy"

Science philosopher William Whew ell would not allow Darwin's book into the Cambridge university library. Charles Thaxton wrote that evidence for design is reasonable "in ordinary life we distinguish natural from intelligent causes all the time, when police officers determine whether a person died of a natural cause or was murdered, when archaeologists decide whether a chipped rock is just a rock or a Palaeolithic tool." Evolutionists have failed to provide fossils to show the natural origins of the species. The Harvard Palaeontologist Dr. Gould wrote, "the extreme rarity of transitional forms in the fossil record persists as a trade secret of palaeontology… all palaeontologists know that the fossil record contains precious little in the way of intermediate forms; transitions between major groups are abrupt. Francis Hitching: "when you look for links between major groups of animals they simply aren't there at least not in enough numbers to put their status beyond doubt. Either they don't exist at all or they are all so rare that endless argument goes on about whether a particular fossil is…transitional between that group or that."

Phillip E. Johnson of the Berkeley wrote, "Darwinism is not so much of an inference from the facts as a deduction from naturalistic philosophy."

In general, models that contradict scientific laws are unscientific. therefore a growing number of scientists have concluded that there is no possible explanation for the complexity of life other than intelligent design. In fact Dr. Granville Swell, states that intelligent design theories are gaining momentum in scientific circles. Therefore, it demonstrates that perhaps slowly but surely scientists are moving forward in the confirmation of revealed religion and the existence of a Creator.

<u>Scientific evidence and belief in God:</u>

Religion and science are often held to be in contrast to one another, each one trying to disprove the other. Actually, they support one another a lot, and religious believers are often happy to find out about scientific proves that support their religion. Science itself is only the study of the physical world, therefore, interpretations and explanations are often correlated with religion and ideology they therefore, cannot be separated. Perhaps they are all paths to knowledge and together explain both the existence of the world and its purpose.

If one knows that there is a Creator it is possible that they created the world however they liked, so perhaps it is possible to find out how the world was really created. If the Creator so wanted we could be created via the process of evolution from monkeys. But we have not found any evidence for that – on the contrary.

All creation, and the human being in the beginning was created from water, we are told in the Quran, and this is what we are taught in science and even from the Bible, that all life began from water. To begin with, we are told that the first human being was created from clay, all human beings on earth take their different skin colours and temperaments and dispositions from the type of clay they are created from, for some are created from dark coloured clay and others from light coloured, and the dispositions of people are different, for some are mild tempered and slow to anger, while quick to forgive and some slow to anger and slow to forgive and some are quick to anger, quick to forgive and some are quick to anger and slow to forgive. We are told that those who are slow to anger quick to forgive are those with the best dispositions. We don't take our colours from the sun or atmosphere, although the atmosphere may make slight changes it can't change the foundation or effect the offspring because skin colour is in the gene pool. So what's the evidence one may ask, how can we know our origin is from soil. There are some simple ways to demonstrate this, the most obvious is that when a person dies we see (or know about) their body decaying and becoming dust before dissolving into the earth. If we were not from the earth we would not be able to return to it completely, there would be something to distinguish us and show that our beginning is from something else, so it would not decay completely, then another evidence throughout human lives is that our body gives off a dust (it's for that reason we must clean our homes regularly) and if a person is thirsty you can see how their skin becomes dry, flaky and dusty, dandruff on the scalp is another evidence and the fact that the human body is susceptible to bacterial and fungal invasions as are plants is another evidence, then there is the fact that our bodies need nutrients from the earth to survive, we depend on nutrients from vegetables and earthly produce for our survival, fruits and vegetables take their nutrients from the soil, and we replenish ours from them. Our bodies are also physically made up of the same amounts of compounds as is soil, for we have the same amounts of nitrates and compounds as soil. We also benefit immensely from using soil to replenish our bodies in the use of clay packs and so forth. We also have similarities to plants who grow out and depend on the soil, some skin diseases even resemble those of plants and we have thin hair or downing on our bodies as do plants we also begin as something small – a seed perhaps, which grows, flowers, flourishes and blossoms before fading withering away and dying. We need water air and sunlight for all of this, so we resemble plants in our physical beginnings, which is an evidence that we are from the earth, our physical selves anyway.

10

THE HUMAN BODY: HOW IT RESEMBLES THE EARTH:

Like everything else on earth the human body is made up of chemicals some of these are solids and some liquids, while still others, gases. Each consists of one or more elements. An element is a fundamental form of matter. No ordinary substance can break it down to other substances. There are ninety two natural elements on earth, and researchers have created other artificial ones. Organisms consist mostly of four elements, : oxygen, carbon, hydrogen, and nitrogen. The human body also contains some calcium, phosphorus, potassium, sulfur, sodium and chlorine, plus trace elements. A trace element is one that makes up less that 0.01 percent of body weight. Trace elements are vital, since red blood cells contain oxygen without the trace element iron. Atoms of elements can combine into molecules the first step in biological organization. Molecules can combine to form larger structure. Many trace elements can be found in human tissue, such as arsenic, selenium and fluorine – are toxic in amounts higher than normal.

The atom is the smallest unit that has the properties of a given element. Atoms are composed of more than one hundred kinds of subatomic substances. Except for hydrogen, atoms have one or more neutrons which have no charge. The most important here are protons, electrons and neutrons.

Scientists can use radioactive decay rates to define very old substances. The human body resembles the structure of the earth in its composition of minerals and elements. This is even more apparent when we know that the human body after death decays into the earth, and had the human body been something other than earth it could not decay completely into the earth unless it was the same structure. Water can be poured into water without leaving any trace. However, vinegar, tea or milk cannot be poured into water exactly without leaving a trace because the components are different. Nothing can dissolve into something else exactly without any trace except if containing the same structure and constituents. Some of the following constituents are shared being the human body and the earth.

Hydrogen, oxygen, carbon, nitrogen, phosphorus, calcium, sulfur, sodium potassium

Chlorine , magnesium, Flourine, ion, zinc, Rupidium, strontium, Bromine, Boran, copper, lithem, lead, cadmium, titanium, cerium, chromium, nickel, manganese, selenium, tin, iodine, arsenic, arsenic, germanium, arsenic germanium, molybdenum, cobalt, celium, mercury, silver, antimony, niobium, barium, gallium, yttrium, lanthanum, Lanthanum, tellurium, scandium, beryllium, indium, muslin – thallium, bismuth vanadium, tantalum, zirconium, Gold, samarium, tungsten, thorium uranium

Fluids inside the human cells is about 7 on the ph scale. Ph of body cells slightly higher – 7.3 and 7.5

"we created man from an extract of clay" Quran 23:12

In an article titled: Was the Bible right about the origins of life? Scientists believe that we may have our beginnings in clay (Daily mail, Daily Mail reporter 2003) suggests that just as the Bible, Quran and even Greek mythology has suggested for thousands of years that life began as earth, dust or clay, the new theory is that clay is a breeding ground for chemicals which it 'absorbs like a sponge'. The process takes billions of years, during which the chemicals react to each other to form proteins, DNA and eventually, living cells, scientists apparently told the journal scientific reports. Biological Engineers from Cornell University's department for Nano scale Science in New York, state believe clay 'might have been the birthplace for life on Earth.' It is a theory dating back thousands of years in many cultures, though perhaps not using the same scientific explanation

"over billions of years, chemicals, confined in those spaces could have carried out the complex reactions that formed proteins, DNA and eventually al the machinery that makes a living cell work."

The conclusions are based on experiments using synthetic hydrogels, adding DNA, amino acids and enzymes and stimulating the production of proteins.

While it may be one theory concerning the creation of life on earth, it may also have modern and moneysaving applications for drug manufacturing. The report added: "why consider clay? Its dirt cheap. Better yet, it turned out unexpectedly that using clay enhanced protein production."

Molecules that make up living things are called biological molecules they are built on atoms on the element carbon. Four classes of biological molecules are carbohydrate, lipids, hydrogen are mainly in the form of water, and carbon makes up more than half of what is left, proteins, and nucleic acid.

Each element of life is an organic compound – containing the element carbon and at least one hydrogen atom.

The human body is mainly oxygen, hydrogen and carbon, and nitrogen. Most carbohydrates consist of carbon, hydrogen, and oxygen atoms. The simple sugar glucose is the main energy source for body cells.

The human body also bears a lot of resemblance to plants. In order for a plant to flower it needs to be nourished and have enough nutrition, sunlight and water for its growth. Human children also need nourishment, care, enough sunlight, water and attention to grow and blossom into young adults. Like plants humans open up and become stronger and fully developed in a certain period of life, known as youth. Proper care in earlier years makes them more open and productive in later years. Like plants, in older years people seem to wilt, their earlier powers have weakened sometimes they have become bent over and lose their colour and powers that they previously had. Flowers as they wither gradually become more dry, and people similarly. Even we have the fine hair on our bodies as we find on the stems of plants. Plants can be subjected to bacteria, funguses and bacteria, and so can people, as well as insects. Perhaps the fact that we resemble plants in such things as well as other animals demonstrates that the human beings are also sourced from the soil although we don't grow out of it directly. Of course human creation is far more that only clay, as has been explained in the Quran, and we are told that actually, as science says the beginning of everything is water. The human being has been created from a blood clot also, and from flesh as is described. The beginning of humanity is created from something small the eye cannot see. When a human baby is born he is equipped with a brain, a nervous system, a heart and veins and arteries and lungs, and hearing and seeing, and that is a miracle that in the beginning he started off as a small egg, or piece of flesh. On our heads we have around 3 thousand hairs and every hair has its own roots, and in eyes there are 10 layers. We are told that the soul does not enter the body and is not male or female until 40 days have passed in the womb. As for the first humans for we are told that the first human was a male, by the name of Adam, and from his ribcage was created his wife Hawwa, and the first humans we are told we not primitive, actually Adam was very learned and knew the names of everything and knew how to do everything, more than the angels and this is one of the reasons humans have an enemy in shaytan, who envies human beings because of this. Adam was not weak he was reportedly much taller than we are currently, and lived much

De-evolution from monkeys and a return to humanity:

longer, living over a thousand years and had many children with hawwa who gave birth many times to twins each time, so it seems that this is opposite of what we are taught in the monkey men story, here we are told that human beings were created whole and complete but we have become smaller in size, live shorter lives and know how to do less (this is especially apparent in our time, where we see all these gajets and gizmos to do things for us, this a big reason for peoples current ignorance of life and laziness).

11
THE SOUL OR SPIRITUAL SELF:

It is obvious that we have a non-physical self, as there is something that leaves the body in sleep or death, and there is something that drives it in our waking lives. Hence no one can ask, why then are we not like a clay pot. A clay pot has not had a soul breathed into it.

The soul of course, cannot be seen, but there is a big difference between a person living and a dead person. We are told that in death, the soul is removed from the body. This implies that in life, it is attached to it, and everything that someone may not be able to do in death, from movement, sight, speaking, thinking feeling, loving hating, disbelieving and believing and so forth are all from the soul (although many things such as seeing as hearing need a connection between body parts to function properly, such as eyes and ears).

Maybe we cannot see the soul, but we see its effects, we also feel it within ourselves. It is commonly thought and believed that the main centre of the soul is the heart, and it is the heart, which perceives, understands, things, believes, loves and hates.

The soul become attached to things. Sometimes people complain of pain or sadness, depression, this is also from the soul, to make a person realise that something is not right, maybe they have done something wrong, or they are not living their life in the right way. People need to care for their soul, as they do for their body.

12:

THE HISTORY AND ORIGIN OF HUMANITY ACCORDING TO ISLAM:

We are told, contrary to popular notions and presumptions that the first and early generations of people, were monotheists worshipping God alone. Then later came people who gradually, with the influence of Satan, began to worship idols, and began to commit crimes, such as murder or crimes stemming from envy and greed. The problem is that perhaps the story of coming from two human beings in the beginning and being created is just so obvious and simple that's it's been taken for granted. People's imagination imagines that perhaps there is more to the story than that, so evolutionary theory or alien invasion stories are told and believed by some. Perhaps they are more interesting for some, after all for centuries people have been saying we were created by God and our origin is from two humans. The problem is when we are told such stories are scientific facts and proven. So what's the evidence that we are from two humans directly without evolving from apes. The evidence is firstly that humanity is made up of two types of human being, the male and the female. The coming together of these two is what brings a human being into existence and keeps humanity going, a man alone cannot create or bring a baby and a woman cannot bring a baby by herself. Each one passes on DNA which contain genes that are passed down and give the foetus certain attributes and characteristics that are found in the parents, such as colouring, height, disposition and so forth, or sometimes not directly from the parents of the child, from relatives that go back further in terms of hereditary, even far back than we are able to trace. This is apparent by the fact that at times a person may give birth to a baby that does not resemble the parents at all, which of course can cause confusion in terms of its parents and fidelity, however, such cases are from genetics. An example, was told to me by my friend from Ethiopia, who told me the story of a black Ethiopian girl who married another black Ethiopian boy, but who gave birth to a child blond haired and blue eyed. Of course people began to speak and accuse her of adultery, for how could a black person give birth to a white baby while being married to a black man? There were no white people in the area however, which made the case even stranger. When the woman had more children they apparently all were as the first child, white skinned, blond haired and blue eyes. It was then that it became apparent that their mother was innocent, and there was no cheating, but these children had inherited some genes from someone in their genetic past – and of cause it must have been someone far back as the parents and grandparents were all black Africans. If the story teaches us anything, it teaches us that humanity is one and we are all human beings, equal with one origin, and we have genetic links with each other. Only members of one species can reproduce together. A monkey and a human cannot have offspring because they are not the same species actually. But humans of any type and colour can marry and have children. Therefore there are no races in reality except the human race. It also demonstrates the importance and reality, and that, contrary to popular thought, children inherit genes, their colouring and so forth, and it is not because of the sun or atmosphere or habitat, which has minimum impact. Similarly a child can inherit from any of their genetic ancestors, who are human and from the same species. If people shared a common ancestor or were evolving, there should be the ability to inherit from previous ancestors, from monkeys or in between stages, or showing signs of babies becoming ever so slowly more developed and evolved. However, such stories demonstrate that genetics is laid out and we are designed and equipped with characteristics, colours and so forth before we even born, which does not support the idea of monkey men slowly developing or anything else. It only shows that two people are needed to keep humanity going and each person can be traced back to two people as long as we trace. The first human would have had to have been created purposefully, for an accident or coincidence cannot bring about the intricate detail and adeptness of a human being.

In fact many things can be traced to their source of they truly existed. In his book the Torah, Quran and Injeel Mauris Baukly shows that the prophets mentioned by religion and found in the Jewish and Christian religion before the revelation of the Quran, can all be traced back, they follow one from another until the reach their source, which is the first human being Adam, who took the first revelation and passed it down to others. How can science explain why there are these men who lived not in one place but in different places, different centuries, and different people and circumstances, all came with the same messages that there is One God, that they are sent as a prophet and that a religious law has been sent to people. Especially since these people did not directly know each other, they did not learn directly from each other and were not appearing to be influenced. Most of them appeared in areas and places where there was polytheism, which was practiced, they and those who believed with them were often attacked. So what made them go to all this trouble to tell people that there is One God? People in our times may hold a belief in One God in their heart, that can be a personal belief, but many don't concern themselves that much about convincing other people. How did these people come with similar laws and beliefs while coming at different times and places? Perhaps religion itself evolved somewhat, one may say, although it began as correct monotheism and from that branched out polytheism and all different types of beliefs, with atheism, as one part of that, for an atheism perhaps has unseen hidden gods or perhaps takes their own self as a deity. At certain points in time it seems that there was an awakening or reminder to people that worshipping statues and wood, animals, stars, sun, or the moon, or fire is not correct and people should worship the Creator of such things. Now a days perhaps we find people worshipping plastic gods. However, there have been frequent reminders throughout time amongst people who were not possible to meet up and make an agreement to tell a big lie to people for fame or some benefit. These people did not ask for money and were known for being honest and upright. How can science and evolutionary theory explain this…especially since some argue that perhaps these were fictional characters that did not really exist. Yet we are able to know the graves and points of many of them, and it is accepted in the Torah as well as the Bible and Quran that such people existed, as it is documented that people such as Socrates, Alexander the Great, King Henry the Eighth and others existed, shall we then deny their existence also to suit us. If we can trace their tombs, places of living, other traces and so forth, we can easily confirm their existence.

Furthermore, the Quran puts people to the test, and tells some stories and makes mention of some people of the past and things that have happened to them. For example, Ad and Thamud, they were destroyed all together in their cities, near Yemen for their disbelief and it is known the areas they lived in and its possible to see the area, in fact the Quran encourages people to travel and to see what happened to previous nations. The people of Lut were also destroyed and the people of the cave who fled from a king of their time trying to persecute them for their religion, were said to have fled to a cave in Jordan. In fact, instead of trying to uncover fossils that don't exist and perhaps do not give correct and accurate information about humanity, it might be more useful for people to put their time and efforts into finding the truth and evidence of these stories that the Quran informs us about. If people are in doubt they are told to explore and bring evidence. The story of Moses and the Pharoah is well known from the Quran and the Torah. This took place in Egypt, in the story we were informed that the Pharaoh was made a sign for people, so his body was preserved and it is said to be in Egypt in the museum there, preserved. Attempts to trace this Pharaoh, who was he exactly, have claimed that he was Ramses. If more study and effort was made to study such stories and uncover evidence, it would be big support and service to humanity, to clarify the truth of the Quran and revealed religions.

CONCLUSION:

To conclude, it seems obvious to the writer that evolution has, and always will be a theory, and as science discovers more intricacies and details of life, it become more apparent that life did not come about by itself, or by accident, rather it must have been designed and so needs a Creator. We may not be able to perceive exactly how our Creator is from our minds alone, but we can know that there is One.

If monkey men theories and aliens can't explain everything, and leave more questions and confusion than answers, there are some beliefs that have withheld the tests of time and cannot be ds-proven, and perhaps are gradually on the way to be proven. Many scientists are now beginning to come around to the idea of a Creator, and acceptance of intelligent design is increases amongst scientific circles. Similarly, science demonstrates that human bodies closely resemble the dust of the earth from, which we are said to have been created from.

How can I know that humans began from two humans and not from apes one may ask. Well, is not every child born from two human parents and they from two human parents and so forth until we reach the beginning of humanity. As I have mentioned, we have not seen any evidence at all for in between stages or small alterations or anything and genes and DNA bring evidence that we are fixed, and we take our abilities and bodies from those before us, we don't come with anything new. Also how could we say that each year we improve and get better and stronger, when actually things in the world and the universe seem to be weakening and becoming less than before, such as our star the sun, which is said to be becoming weaker, we now benefit less from the earth, as it contains fewer nutrients and more foreign things such as chemicals and so forth, meat contains hormones. Ice caps in Icelandic countries are now melting and heat is increasing in colder climates, now if babies are born with any mutations or differences is malfunctions, or nutrient deficiencies, brain problems, autism, and effects of nuclear radiation and global warming. It would seem that in our current time people don't seem to be at the same intellectual standards of those before, or morally. There was not this level of killing, fighting, stealing, promiscuity or crimes previous to this, although they were in existence, and there were other issues. So if anything perhaps we have digressed – both morally and physically and even mentally and spiritually, although there is the use of technology to speed things up for us, no doubt making up for our shorter lives. But saying this it seems that that there have always been and still are people who live until 100 and perhaps a little more or a little less and perhaps the average life span of people is dependent on certain times and certain lifestyles, as well as from our genes. It is also apparent from now that a major difference between past acceptance of creeds were less able to be tried and tested, now there is equipment and the ability to learn and study the world in depth more and to bring more concrete evidences for ideas and beliefs. This does not mean however, that everything before was all wrong and does not mean religion is outdated. It may be that in these times, we may be more able to prove its authenticity. This is because the truth, wherever and whatever it is, can tolerate analysis and remains after scrutiny and study and can bring arguments and evidences, whereas what is not true cannot. Also what is correct should bring peace and harmony amongst people and not turn them away from their humanity. It should improve people's lives, and the relationships between people. For some time these have been waning in British society. Now I felt for some time, that in Britain, humanity has weakened, could it be that we are losing our humanity?

This brings me to the last point, which is that a return to humanity and humanness can only begin if we are able to return to the correct understanding of what a human being is exactly and where they are from. Ape people may lead to ape like behaviour, after all their ancestors were doing these things thousands of years ago right? So what's the problem? Eating bananas and swinging from the light fittings are quite normal then. But what about if we were all human being from the same beginning? Then we are all the same humanity and all equal. There are no better or worse except in behaviours and what a person chooses to do or how they want to live their lives. We are all sisters and brothers, all eating similar foods, and having similar thoughts feelings and needs and the potential for intellectual or social pursuits. Perhaps it is the intellect and religion that separates a human from an animal. And how are we to explain all these spiritual sensations and experiences scientifically if we are barred from their research? If any person is a true scientist they will research and study such things objectively – meaning that if they come to the conclusion that the soul or God exists from that, they accept that, if not then they are treacherous and not doing justice, to learning, science, society or humanity or at large, and a scientist also should have manners and etiquette when dealing with others. Is calling others stupid, backward or so forth, scientific and progress? No, so these are some of the issues that I wanted to bring about, as a student of knowledge.

REFERENCES:

Bucaille, M., *(2003) The Bible, Quran and Modern Science*

Bacaille, M., *The Torah, Quran and Bible*

Behe, M, J., *Darwin's Black Box* (1996)

Bainbridge, D., (2009) *It's not teenage strop. It's a key part of evolution.* The Times

Carnie, D., *(1990) Stop Worrying, Start Living*

Davidson, S, Dr., Morgan, B., (2002) *Human Body Revealed* A Dorling Kindersley book

Denton, M., (1986) *Evolution: A theory in crisis* Harper and Row

Dossey, L., (2003) *Healing Beyond the Body,* Shambald

Ernst Haeckel en.wikipedia.org

Elio Frottoli, (2002) *Healing the Soul in the Age of the Brain*

Embryology (2000AD) Muslim World League Leading Press, Saudi Arabia, Mecca.

Encyclopedia Britannia

Gould, S, J. (1979) *a quahog is a quahog*

Gould, S, J. (1980) *Is there a new and general theory of evolution*

Gould, S, J. (1977) *the return of hopeful monsters*

Hassani, S, TS., *Muslim Heritage in Our World* (1001 Inventions) second edition (2006)

Hawkes, J, H., (1963) *The World of the Past*

Hawkes, K., O'Neil, JF., Jones, NG., (1991) *Hunting income patterns among the Hadza: big game, common goods, foraging goals, and the evolution of the human* www.ncbi-Bnim.nih.gov

Insel, P, M., Roth, W, T., *Core Concepts in Health* (2002)

Imber, D., (1970) *Faked fossils of primitive man* www2clarku.educ

Jabr, F., (2010) *The evolution of emotion: Darwin's little known psychology experiment* scientificamerican.com

Kathir, Ibn., *Explanation of the Quran*

Kausar, N., (1976) *The creation of man* Sh. Muhammad Ashraf

Leakey, R., *The challenger in dispute of evolution* en.wikipedia

LeDoux, J, E., (2013) *Evolution of human emotion* www.ncbi.nlm.gov

Mahmud, M., (2000) *Dialogue with an atheist* Dar al-Taqwa

Martin, S., (2007) *shock claim: Human DNA 'was designed by aliens'*

Mayer, S., (2004) *Darwin's Doubt: the explosive origin of human life and the case for intelligent design*

Murphy, N., (2006) *Bodies and Souls or spirited bodies?* Cambridge university press

The Natural History Book (2010) Dorling Kindersley Limited

Nawawi, *Forty Hadiths*

Persuad, M., *The Developing Human: Clinically Oriented Embryology* (1998)

Persinger and Cook, (1998) Experimental induction of the 'sensed presence' www.science-frontiers.com

Rice, S., Neurological link between epilepsy and religious experiences (University of Missouri Columbia) Neurosciencenews.com

Robinson, J., McCormick, D, J., (2011) *Concepts in Health and Wellness Concepts in Health and Wellness*

Roberts, A., Dr., *The Complete Human Body: The Definitive visual guide* (2016) DK enhanced.

Starr C., McMillan, *Human Biology* (2010)

T.V.N, *Human Development as Described in the Quran and Sunnah: Correlation with Modern*

Vogel, A., *The Nature Doctor: a Manual for Traditional and Complementary Medicine* (1991)

Wenz, J., (2017) *Skull of Homo erectus throws story of human evolution into disarray*

Winston, R., *Human* (2004)

Woodward, T., (2003) Doubts about Darwin: A History of Intellectual design

Zindani, A., Johnson, E, M., Goeringer, G, C., Simpson., Moore, K, L., Ahmed, M, A., Persuad,

Zimmer, C,. *(2007) The Descent of Man*

De-evolution from monkeys and a return to humanity:

ABOUT THE AUTHOR

The author is a British born Muslim woman who converted to Islam at the age of seventeen after learning about it through friends. She completed a degree in psychology and history of ideas and then went on to complete a degree in the fundamentals of Islam and studied hadith and explanation of the Quran. She has also has interest in science, natural health and nutrition and real life stories and issues.